Mazal Tov from the PTA
on your Graduation
from
The JEC High School

June 2023
5783

The Lazarus Edition

Let There Be TIME

The Meaning and Messages of the Jewish Months

SHAYA WINIARZ

Let There Be

ISRAEL BOOKSHOP
Publications

The Lazarus Edition

TIME

The Meaning and Messages
of the Jewish Months

SHAYA WINIARZ

ISBN 978-1-60091-674-8

To reach the author please email:
shayawiniarz@gmail.com

Book & cover design by:
VIVIDESIGN
SRULY PERL | 845.694.7186
mechelp@gmail.com

Proofreading by E.M. Sonenblick
esonenblick@gmail.com

Published and distributed by:
Israel Bookshop Publications
501 Prospect Street
Lakewood, NJ 08701
Tel: (732) 901-3009
Fax: (732) 901-4012
www.israelbookshoppublications.com
info@israelbookshoppublications.com

Printed in the USA

Distributed in Israel by:
Tfutza Publications
P.O.B. 50036
Beitar Illit 90500
972-2-650-9400

Distributed in Europe by:
Lehmanns
Unit E Viking Industrial Park
Rolling Mill Road,
Jarrow, Tyne & Wear NE32 3DP
44-191-406-0842

Distributed in Australia by:
Gold's Book and Gift Company
3-13 William Street
Balaclava 3183
613-9527-8775

Distributed in South Africa by:
Kollel Bookshop
Ivy Common
107 William Road, Norwood
Johannesburg 2192
27-11-728-1822

Rabbi Reuven Feinstein
131 Bloomingdale Road
Staten Island, New York 10309
718-317-0819

שלו׳ ראובן פיינשטיין
ראש הישיבה
ישיבה ד׳סטעטן איילענד
718-356-4323

I was asked to write an approbation for a sefer by Shaya Winiarz.

For lack of time, I have not been able to read through the sefer, but I am familiar with its topic as well as its author.

The topic is Segulas Hazman – how we must serve Hashem differently throughout the different months of the year. This is a *chashuv'e inyan*, and one that is often overlooked. It is important for Bnei Yisrael to learn this topic, and I would therefore like to commend Shaya Winiarz for putting in considerable time and effort to make this topic readily available to the English-speaking public. The source material is difficult, and it is a great accomplishment to distill it and present it in a clear fashion.

I know the author, Shaya Winiarz, for a number of years. He is a ben Torah and I can vouch for him. He is serious about his Torah learning, and I am confident that he has succeeded in presenting Toras Emes is this sefer. His writings should be studied and implemented by Yidden worldwide.

May Hashem bless the author with much success in spreading Hashem's ways.

לזכרון עולם

This *sefer* is dedicated in memory of my grandparents, Rabbi Yosef Dov and Chana Fraida Lazarus. My siblings and I all feel that we were blessed to have had such wonderful people in our lives. Each in their own way, they left an indelible impression on us.

Zaidy was a gentle, soft-spoken man. His children and grandchildren attest that one would never hear a word of criticism from him. He was a father and a teacher, and in both the school setting and at home, he taught primarily by example rather than through verbal instruction. He wouldn't say that *lashon hara* is wrong, he would merely walk away if someone began speaking ill of others. He wouldn't tell people how precious Torah is, he would spend all his days learning in full view of others – be it at a wedding, on line in a store, or during any spare time that he had. He wouldn't preach that time is precious, he would carry a *sefer* with him wherever he went so that no minutes would go to waste.

Zaidy loved learning all branches of Torah, and had a particular love for Tanach. He also loved Eretz Yisrael. We knew that we could always play a game with Zaidy in which we would show him a picture of a place in Eretz Yisrael, and he would be able to identify it with ease. He would then tell us where in Tanach this place was mentioned, and what significance it has to the Jewish People. Zaidy also taught himself Hebrew at a young age, and enjoyed speaking it with others whenever he had such an opportunity.

Zaidy also enjoyed photography. He pursued this hobby throughout the different technological eras in which he lived, progressing from slides to film to digital files. He would patiently read through the manuals (!)

and master each one. Zaidy would photograph us – his grandchildren – in all sorts of creative outdoor settings, thereby combining his love of family and photography with another one of his loves – nature. Zaidy had a green thumb, and would grow a beautiful garden every year in his backyard. In Zaidy's garden one could find a few *esrog* trees, an apple tree, as well as tomatoes, squash, cucumbers, peppers, string beans, cabbage, and more. Zaidy also transplanted lilacs from his childhood home in Spring Valley to his house in Connecticut, and then transplanted them to his next home in Staten Island when he moved there. I clearly recall Zaidy telling me how he knew that different plants needed different amounts of water based on what the Gemara says.

Zaidy was a consummate *mechanech* (educator), and taught for many years in yeshivah elementary schools. In addition to writing curriculums, Zaidy also would use his beautiful handwriting to design elaborate picture worksheets to engage his students. His calligraphic skill also enabled his side occupation of checking and fixing *STAM* (*Sifrei Torah, Tefillin*, and *Mezuzos*). Zaidy also taught adults in his capacity as a *maggid shiur* for the Daf Yomi. I recall Zaidy telling me that he would always carry a Gemara with English translations, so that no individual would feel that Gemara study was out of reach for him. After all, if the rabbi could use an English Gemara, so could he. In a similar vein, when Zaidy was asked why he – a rabbi – never grew a beard, he responded that he didn't want people to feel that he was any different than them; anything he was doing, they could also achieve.

When his hearing deteriorated and he could no longer teach, he and Bubby opened a *sefarim*/Judaica store. Zaidy was a bibliophile – a book collector. He loved finding new *sefarim* (and learning them!). Zaidy's library in his home had over three thousand *sefarim*, and most of them had his writings in them – annotations, comments, and corrections. The *sefarim* store fit with Zaidy's love of *sefarim*, and was an ideal way for him to stay in the world of *kedushah* (holiness) even after formally leaving the world of *chinuch*. Zaidy worked hard in the store, but once a customer walked in, he became fairly indifferent to making a sale. What he really wanted was to explain Judaism or discuss a *sefer* with the customer. He was never lacking a *dvar Torah* to share, and brought many people closer to Hashem from his position behind the counter.

Bubby was also such a fixture in our lives. She loved having us around, be it in her house or in the *sefarim*/Judaica store, and we loved being around her. And whether we went to see her in the house or in the store, we would never walk away empty-handed. Bubby loved giving, and we – as grandchildren – knew to be careful in what we expressed interest, because our expression of interest usually resulted in Bubby saying, "Okay, so take it home with you!"

Bubby cared about others and would always think of other people. She would remember everybody's birthday, and would send cards to everyone in honor of their occasion. Bubby never forgot a card, because she never forgot those she cared about. She was always thinking about her loved ones. Bubby would keep a list of people who she would always *daven* for, whether they needed *parnassah*, a *shidduch*, a *refuah*, or anything else. When one would visit Bubby in her house, one could very reasonably expect to find her on the couch, saying Tehillim.

Bubby would always recognize and acknowledge all of her blessings. A byline of hers was, "I'm so lucky."

Bubby was a strong lady, and had to have been to have raised four very rambunctious children. She was tough, yet she was able to encourage others. She had a good word to say about most, and in the other situations she would simply say nothing. Bubby had an admirable skill to seal her lips and say nothing rather than badmouth others.

She also had a love of a good laugh. Bubby was able to tell or listen to a funny story or joke which would set her off with a hysterical peal of laughter that forced everyone around her to join in.

For all this and more, Zaidy and Bubby were precious figures in our lives and the lives of many. May their memory be for a blessing.

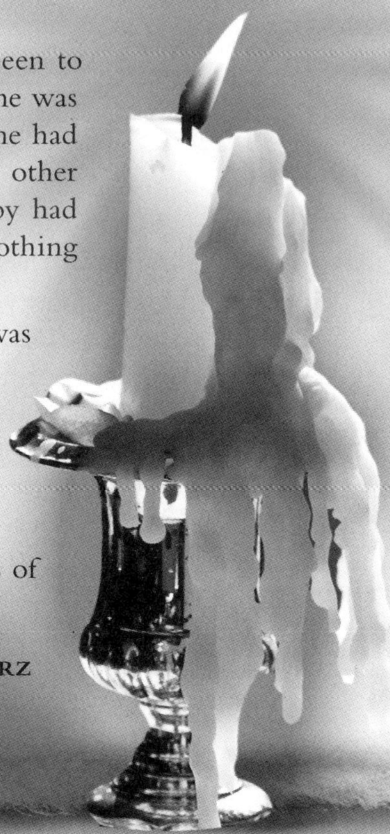

Shaya Winiarz

לעילוי נשמת

הרב **יוסף דוב** בן ר' **יעקב אליעזר**

חנה פרידא בת הרב **ישעי'**

Our parents had an admirable ability to stay focused on their life goals. They were wholly dedicated to building a Torah-true home where good *middos* and integrity would be cultivated. In a world where raising children to be good, kind-hearted adults was by no means assured, our parents achieved greatness. Their ability to love and embrace knew no bounds. Our mother was a determined lady, a powerful guiding force. Our father was gentle and kind – a scholar, teacher, and righteous man. Their parenting approach was not flashy, but it was effective. It has often been said (because it is true), *Rabbi Yosef and Channie Lazarus led by example.* Learn. Teach. Serve others. Never speak badly of anyone. Never give up on anyone. Forgive quickly. Be grateful and laugh often.

Children cannot fully thank their parents – their sacrifices are too great and their gifts too precious. What we do instead is stay true to their values and hope to make them proud. We dedicate this wonderful *sefer* in memory of our parents. Their holy souls shine brightly, illuminating the path forward for all who knew them.

Miriam (Lazarus) Winiarz

Yacov and Pamela Lazarus

Yehuda Lazarus

Moshe Lazarus

In Memory of
Channa (Spiegel) Lazarus

Channie was that special person that cared for everyone.

She loved her family with a full heart and gave to all unselfishly and with joy.

She was so proud of her children and grandchildren.

It was Channie who brought Rabbi Shaya Winiarz to my attention with his wonderful articles for *The Jewish Press*.

May the *neshamos* of both Channie and Rabbi Joe Lazarus have an *aliyah* with the production of this book of Torah thoughts.

Naomi Klass Mauer,
Publisher of The Jewish Press

Rabbi Yosef Dov Lazarus with his parents
Louis and Frieda Lazarus

לעילוי נשמת

ר׳ יעקב אליעזר בן ר׳ יוסף דוב

פריידה בת ר׳ משה

Our grandparents, Louis and Frieda Lazarus, did not come from what one might call "a yeshivish background." Yet, they both recognized the importance of Torah and a life of being actively involved in Torah study.

They were dedicated to this ideal and were committed to infusing their lives and the lives of their children with Torah. Many times, they made choices with specific intention to move toward Torah, both in their family and community.

They were successful in this goal, and their efforts brought forth children and grandchildren who live lives steeped in Torah study and education. The credit they deserve for generations of Torah-true Jews cannot be overstated.

Miriam (Lazarus) Winiarz
Yacov and Pamela Lazarus
Yehuda Lazarus
Moshe Lazarus

Table of Contents

Tammuz

Av

Elul

Tishrei

Mar-Cheshvan

Kislev

Teves

Shevat

Adar

Introduction

Why do we need an introduction?

In my opinion, a *sefer* of this type needs an introduction. This is because the topic covered herein is not commonly included in the curricula of most Torah institutions around the world. Presumably, the reason for this is simply because there is limited time and ability granted to both the teachers and students. Consequently, this topic—as well as innumerable other branches of Hashem's Torah—is overlooked. One should not think, however, that only the topics that we are taught in school are important. When I once asked a *rebbi* of mine if the *yeshivos* omit certain branches of the Torah because they view them as unimportant, he responded, "Of course we think they are important! But you only have a few years in yeshivah. If we teach those areas, you will only know those areas. But if we spend all our time on Gemara—which is the most complex and important part of our Torah—then you will gain the most important part of Torah, as well as the tools needed to move on to all those other topics on your own."

With this in mind, I present a *sefer* on a branch of Hashem's Torah that is not commonly studied in depth. It's a topic I find interesting and that I wrote a newspaper column on for a few years. My work within this book is in no way an exhaustive treatment of the topic, but it's a good start.

What is the topic of this *sefer*?

The topic is that of the Jewish calendar. We are not referring to the mathematical computations and formulas that go into the extraordinarily complex topic of arranging our calendar in the proper halachic fashion. Rather, we will be discussing *Segulas Hazman*. This is a topic alluded to throughout *Chazal* (i.e. Talmud, Midrash, etc.)

and elaborated on extensively in later works including, but not limited to, the *Sefer Yetzirah*, the Arizal, the *Bnei Yissaschar*, the Ramchal, and *Pri Tzaddik*. These latter sources are the primary ones from which I draw my ideas, but this topic is discussed by other commentators as well.

While this *sefer* does indeed contain *answers* that I originated, I try very hard (repeat, very hard) to not originate any *ideas*. That is to say, while I may not know for certain whether I divined the correct answer to a particular question, I will not suggest an answer unless I know that the concept contained in the answer is true and sourced in our Torah literature.

So what exactly does *Segulas Hazman* mean? Simply put, it's the idea that not all time periods in the year are the same. The Jewish calendar is divided up into many different parts, each with its own unique characteristics, opportunities, and responsibilities. One can easily understand, for example, that there is a "Pesach season" in the spring, during which we prepare for and observe the holiday recalling the Exodus. In this season we have a responsibility to clean all the *chametz* (leaven) out of the house, and we have the opportunity to fulfill the mitzvos of matzah, *maror*, and the Four Cups. Similarly, we know there is a "Yamim Nora'im season" during which we must do *teshuvah* (repentance), and we have the opportunity to coronate Hashem as our King. Similar "seasons" exist throughout the year.

But the truth is, *Segulas Hazman* is so much more than simply having different mitzvos during different parts of the year. If it were only that, then one could theoretically move all the calendar observances around. We could observe Sukkos in Nissan, Yom Kippur in Kislev, and Chanukah in Tishrei. Why do we observe each Yom Tov at the time that we do? Obviously, the answer is because Hashem told us to. But why did He set it up this way? The answer lies in the deeper understanding of *Segulas Hazman*. For our purposes, we will divide the following exposition into twelve steps. (I kind of thought that would be an appropriate number for this topic.) The layman can read the basic text of this introduction; the scholar can pursue the sources and elaborations found in the footnotes.

I. Making time for us.

Let's introduce the idea that time is a creation. Just like Hashem created the world and He created Man, He also created time.[1] Each day is a creation made by Hashem in order for mankind to "have the time," so to speak, in which to serve Him and earn the eternal reward of the World to Come. Of course, if time is a creation, it is not infinite or eternal. Hashem only created a limited amount of time. To be specific, our world is created with 6,000 years of time.[2] Since there are 6,000 years in our world, if we would assume that there are 365 days in a year, that would mean Hashem created a grand total of 2,190,000 days for Mankind to work and earn the eternal world.[3]

II. Same old, same old?

According to the above, one might assume that all the days that Hashem created are the same. That is to say, there are over two million days and each one, like the one that preceded it, is just one more step until time is up. However, this worldview is incorrect. The Ramchal explains (*Derech Hashem*, part 4, ch. 4): *Every single day of the six thousand years is itself necessary, by Divine decree, having its own*

---

1. This idea is found in innumerable sources, primarily the *Sefer Yetzirah*, which explains that the three roots of creation are *olam*, *shanah*, and *nefesh*—the world (not including man), time, and man. To put these three creations together, one could say man is the purpose of creation and he is placed under the limitations of the other two creations—time and space. The Written Torah even alludes to the creation of *zman* in the verse, *sheishes yamim asah Hashem*—"six days Hashem created" (*Shemos* 31:17). It doesn't say "*in* six days Hashem created," but rather "six days Hashem created" (*Pri Tzaddik, Kedushas Shabbos, Ma'amar* 2).

2. This is explicitly stated in the Gemara in *Sanhedrin* 97a and *Avodah Zarah* 9a.

3. The Ramchal writes this idea in *Derech Hashem* (part 1, ch. 3): *G-d's goodness decreed that there be a limit to man's effort required to attain perfection. After this period of effort is completed, he attains his level of perfection and is then allowed to enjoy it for all eternity. G-d therefore created two distinct periods, one as a time of earning, and the other as a time of receiving reward. [However,] that which pertains to good is always greater. [Therefore,] the time period for earning is limited, and lasts no longer than G-d's wisdom decreed suitable for His purpose. The period of reward, on the other hand, has no limit, and man continues to derive pleasure from his earned perfection for all eternity.*

illuminations, influences, and states, all being essential to complete the world cycle and attain the goal of ultimate perfection. Each day is literally a new creation, adding to and vitalizing the rest of creation. This means that Hashem has a grand plan for the world, and each one of the 2,190,000 days is contributing an essential piece to making this vision a reality.

This would mean that even if you spent fifty years doing the same mitzvos every day, it would not be "same old, same old" to continue doing those mitzvos for another fifty years. The fact that you contributed yesterday's piece of the puzzle doesn't mean that it's not beneficial to contribute a *different* piece of the puzzle today. Similarly, the Chafetz Chaim in *Ahavas Chessed* (part 2, ch. 8) quotes the *Zohar* as saying that the spiritual manifestation of each day will accompany you to the Heavenly Tribunal and testify on how well you did during that day. The Chafetz Chaim here is teaching that just because you studied Torah and did acts of kindness all day yesterday doesn't mean that today doesn't need that vital infusion of spirituality as well. And just because Shema was recited yesterday doesn't negate the essentiality of today's Shema in the grand scheme of the ultimate puzzle.

III. Linear or cyclical?

We've already learned a lot about *zman*—how it's a creation that Hashem grants mankind to enable us to work toward Hashem's goal and thereby achieve eternal reward. We also learned that each day is bringing our world one step closer to perfection. However, if we would stop here, we would not yet have the idea on which this *sefer* is based. This is because one could have said all that we've said and still have assumed that time is linear. Meaning, one can assume that each day follows the other and that once a day fades into the past, it no longer relates in any way to the days we are now experiencing in the present.

Why is this idea untrue? Without a time machine, isn't it true that "then was then" and "now is now"? The answer is—no. According to the truth revealed to us by the Torah, time is actually cyclical. This is true in many ways, so let's take it slowly. Here's an easy example: Even a secular person understands the concept of a weekly cycle. The first day of the week is Sunday, the second day Monday, then Tuesday,

Wednesday, Thursday, Friday, and Saturday. What comes after Saturday? Why, Sunday of course! The eighth day is not a new day in an unending line, but rather it is once again the first day of the week (albeit a new week). Similarly, the first day of the month is called September first, then comes the second and then the third, and so on until day thirty. Then, when the month is complete and we begin a new month (October), we start again from the first, then the second, and so on. We don't continue counting ad infinitum. To give a third and final example, when we complete the twelve months of the year, instead of going on to a thirteenth month, we start again from the first month. Thus, we see that even in the secular calendar there are cycles—weekly, monthly, yearly, etc. Time is not a line but rather a circle.

IV. Days and months of the cycle that match are similar.

Let's take it a step further. It is well known that Sunday is a great day. Everyone is happy that they are off from work. Conversely, people groan when they think of Mondays, because it's the day that they must go back to their difficult workweek. What this means is that Sunday has unique characteristics particular to Sunday, and these characteristics determine how we act and react when this day comes, and the same goes for Monday. This cycle repeats itself every week. Similarly, if one gets paid on the fifteenth of the month, the fifteenth is a happy day for him—a day that comes back again during the next thirty-day cycle.

So now let's apply this idea to the Jewish calendar. Hashem created each day of the week to be unique and different from the others. That means that each Sunday has its own characteristics that make it differ from Monday, and Monday has its own characteristics that make it differ from Tuesday. (The fact that we recite a different Song of the Day at the end of our morning prayers each day is indicative of these unique qualities.) Additionally, besides for the differences between the days of the week, each month in the yearly cycle has its own unique characteristics. Hashem created Nissan with its own qualities, as well as Iyar and Sivan. So while Sivan in the year 5778 provides the opportunity to contribute pieces of the grand puzzle that cannot be contributed during Sivan of the year 5779, the *ways* in which we create those puzzle pieces during Sivan are always same—whether in 5778 or 5779. This is because all Sivans are alike.

V. Let's be more specific.

Hopefully, the concept of *Segulas Hazman* is becoming clearer, but perhaps it will be easier to understand if we give specific examples. Let's take the month of Nissan as an example. The Gemara (*Rosh Hashanah* 11a) teaches*: In Nissan we were redeemed (from Egypt), and in Nissan we will be redeemed (from the final exile).* Now, let's ponder—why is it that both the past and the future redemptions take place during Nissan? Is the Gemara merely telling us a cute coincidence, that they will both "happen" to fall out in the same month? Obviously not. Rather, the Gemara is alluding to the fact that Nissan is a month that is *mesugal* for redemption; that is to say, the unique qualities of the month are related to redemption. Therefore, it is only natural that both redemptions will take place in the month most appropriate for it.

A line of reasoning we will follow in this *sefer* is trying to discover *why* the month of Nissan is *mesugal* for redemption. In addition to this intellectual pursuit of truth, we will also try to discover how we should actively be serving Hashem during this month. The reason for this is twofold: First, although we should always serve Hashem in as many ways as we can, it is appropriate to serve Him in the particular ways unique to each month.[4] In this way, we compartmentalize all the different areas of growth into different parts of the year. Secondly, a person will succeed in his attempts to grow in a particular area much more when the month is *mesugal* for that area of growth.[5]

VI. What can make a month unique?

Now here comes the fascinating question—what makes a month unique? In what ways did Hashem create the months different from

---

4. This idea is contained in every *ma'amar* of the *Pri Tzaddik* on the months.

5. Not only is this obvious, but see Ramban in *Bereishis* 1:18 and *Shemos* 12:3 where he discusses how the month's *mazal* influences and "rules" over that month (while obviously Hashem is above all of them—see Ramban, *Shemos* 6:3).

one another? There are many answers to this question. There are twelve months in the year, and each month's uniqueness is represented by what it aligns with from the following sets of twelve. [See Appendix, where these ideas are organized into a chart.]

- The Jewish people are divided into twelve tribes—each with its own unique capabilities and responsibilities to the nation. Each month of the year is aligned with one of the tribes.[6]
- There are twelve readily visible constellations (*mazalos*) that lie on the path taken by the sun as it travels across the celestial sphere over the course of a year.[7] Each month of the Jewish calendar is aligned with one of these *mazalos*.[8]
- There are twelve letters of the *aleph-beis* that are called *osiyos peshutos*.[9] Each month of the year corresponds to a different one of these letters.

---

6. This idea is found in a number of sources in *Chazal*. For starters, there is the *Midrash Tanchuma* in *Vayechi* (sec. 15 or 19, depending on the edition). It is also found in the *Midrash Yalkut Shimoni* on *Yeshayah* (470). The Ramban (*Devarim* 33:6) elaborates on this idea and sources it in the *Sefer Yetzirah* (5:2), the *Sefer Habahir* (95), as well as a Gemara in *Brachos* 32b (although not all editions of the Gemara contain the Ramban's version). This idea is also discussed extensively by the Arizal. Although there are those who differ as to how to align the months with the tribes (such as the Raavad and the *Midrash Peliah*), we will follow the alignment favored by the Arizal and the Vilna Gaon (*k'seder hadegalim*—in the order that the tribes camped in the desert), which seems to be the most accepted.

7. This idea can be observed even physically with the following understanding: As our planet rotates on its axis, the twelve constellations of the Zodiac rise in the east one by one, at two-hour intervals, and set in the west. During Nissan, the *mazal* of Aries rises first at dawn over the eastern horizon. The Ramban in *Bereishis* (1:18) explains that, as a rule: *The* mazal *that rises [first] by day rules over that day.* Meaning, the *mazal* of Aries, which rises first during the month of Nissan, will exert the most influence on Hashem's world during Nissan.

8. This too is discussed innumerable times in *Sefer Yetzirah*, *Pri Tzaddik*, *Bnei Yissaschar*, and in the teachings of the Arizal.

9. The *Sefer Yetzirah* (ch. 1) explains that Hashem created the world with the twenty-two letters of the *aleph-beis*. The letters are subdivided into three groupings of three, seven, and twelve letters. The group of three is called *Imos*, and are the letters *aleph*, *mem*, and *shin*. The group of seven is called *Kefulos* (lit. doubled) because they can

- There are twelve *pe'ulos* (perhaps we can translate this as "faculties") that we make use of (e.g. sight, hearing, smell, etc.), and each month is aligned with one of them.[10]
- The *Sefer Yetzirah* teaches that the body can be viewed as having twelve prime parts,[11] and each month aligns with one of them.[12]
- There are six parts of the head, and the number is doubled, as we count them for a man and for a woman.[13]
- The four letters of Hashem's Name can be rearranged into twelve different permutations, and each month aligns with one of them.[14]
- There are twelve *alachsonim* (diagonal borders), which are the *zro'os olam* (arms of the world).[15]

Now, obviously this type of discussion is very esoteric and heavily based on Kabbalah. Therefore, since this is not my area of expertise (and presumably, dear reader, it is not yours either), we will steer clear of attempting to explain these deep concepts.

However, what we *can* learn from here is that each month has its

---

be hard—with a *dagesh*, or soft—without a *dagesh*. The *Kefulos* are *beis*, *gimmel*, *dalet*, *kaf*, *pey*, *reish*, and *taf*. The remaining group of twelve letters is called *Peshutos* (lit. simple ones) because they do not change.

10. The *Sefer Yetzirah* (ch. 5) lists all twelve: *sichah*, *hirhur*, *hiluch*, *re'iyah*, *shmi'ah*, *ma'aseh*, *tashmish*, *rei'ach*, *shinah*, *rogez*, *le'itah*, and *schok*.

11. *Sefer Yetzirah* calls these the twelve *manhigim b'nefesh zachar u'nekeivah*.

12. The twelve are right hand, left hand, right foot, left foot, right kidney, left kidney, gallbladder, small intestine, liver, *kurkiban*, *keivah*, and spleen. (*Keivah* and *kurkiban* are both parts related to the stomach that do not have exact translations into English that I know of.)

13. This idea is expressed by the Arizal in *Pri Eitz Chaim* (*Sha'ar Rosh Hashanah*, ch. 4) and *Likutei Torah Ta'amei Hamitzvos* (*Parshas Bo*). They are: the skull, two ears, two eyes, the nose—all of a woman. Then the same six, but for a man.

14. This idea is found in the teachings of the Arizal quoted in *Mishnas Chassidim, Nissan*, ch. 1. There the Arizal writes that when we say Hashem's Name in the middle *brachah* of *Shemoneh Esrei* of Rosh Chodesh *Mussaf*, we should have *kavanah* (intent) regarding the particular permutation of that month.

15. See Ramban (*Devarim* 33:6), quoting *Chazal*.

own unique characteristics. The articles and essays you will find within this *sefer* will focus on certain qualities related to each month. Some essays will work to explain why the month in question is aligned with its tribe, while others will elaborate on the letter, *mazal*, or *pe'ulah* of the month. As a rule, I have steered completely clear of any discussion regarding permutations of the Divine Name, *alachsonim*, parts of the head, or any of the other deeply Kabbalistic topics far beyond the scope of my abilities. Our focus will be *avodah*-oriented—learning from the months' characteristics about the different areas in which we should grow. This idea is what we are bringing out in this introduction—that every time, for example, Adar comes around, we should try to grow in the ways unique to Adar. Additionally, we will succeed in growing in the Adar areas during that month more so than during any other month!

VII. Viewing the names of the months as guideposts.

Fascinatingly, the *Pri Tzaddik* and *Bnei Yissaschar* both add the following dimension to our discussion (although we admittedly won't be analyzing this dimension too often in this *sefer*). They teach (based on *Chazal* and Kabbalah) that the names of the months hint to their unique characteristics. For example, the name Nissan is cognate with the Hebrew word "*nes*"—miracle. The implication of this is that Nissan is a month that is especially primed for miracles (as we alluded to earlier in V.). Another example can be found in the *Midrash Vayikra Rabbah* (29:8), which discusses the meaning of the name Tishrei. The *Pri Tzaddik* also elaborates on the meaning of the name Kislev in his *ma'amar* on that month.

Incidentally, the *Pri Tzaddik* and *Bnei Yissaschar* seem to differ as to the precise explanation for this phenomenon of the names as hints. *Pri Tzaddik* (*Ma'amar L'Rosh Chodesh Kislev*) says that when the Beis Hamikdash was destroyed, the spiritual antennae of the Jewish people became weakened. Consequently, we lost the ability to feel the uniqueness of each month on our own. Therefore, the *Anshei Knesses Hagedolah* established names for the months to hint at our unique *avodah* (specific job) during that month. Up until that point, our months were merely numbered—month 1, month 2, etc. But in order to help us in our post-Mikdash state, the Sages provided us with guideposts—the names of the months.

Bnei Yissaschar, on the other hand, says (in *Ma'amar L'Chodesh Nissan* 1:7) that the names of the months were received at Har Sinai. They were initially classified as part of the Oral Torah, but the Sages in Bavel subsequently allowed it to be incorporated into the [parts of the] Written Torah [which were subsequently written—like *Megillas Esther*].[16]

VIII. Is it only based on how Hashem created them?

Up until now we have been dealing with the idea that the months are unique because that is how Hashem made them. We explained that from the time of Creation, Hashem formed each month with its own unique qualities. However, this is not the only reason the months are unique. The Ramchal in *Derech Hashem* (part 4, ch. 7) explains that *zman* can be affected by what occurs to human beings long after creation as well. Let's take a look:

Besides for the holiness that [intrinsically] exists to varying degrees [because Hashem invested that holiness in them]...[there is also the idea that] on each of these special [holidays] something happened [in history] whereby a great rectification was accomplished and a great Light shone. The Highest Wisdom decreed that on every anniversary of this period, a counterpart of the original Light should shine, and the results of the rectification renewed to all those who accept it.

Pri Tzaddik (*Ma'amar L'Tu B'Av*) also brings this idea: *All holidays can be explained in the following way: A past occurrence happened on that day that made an impact on that calendar date. A degree of holiness was brought into that date forevermore and it is therefore appropriate to make a celebration (i.e. holiday) on that day.*

From these pieces we can learn even more about the uniqueness of the months. Each month was created differently, true. But each month is also affected by that which occurs to our nation throughout the

16. Then again, perhaps the *Pri Tzaddik* and *Bnei Yissaschar* are not really arguing. Perhaps the enactment referred to by the *Pri Tzaddik* was the incorporation of the Masoretic names...

millennia of our existence. So even though the months were created unique and distinct from one another, the Jewish people can also impact the months and contribute to their essential natures (i.e. of the months). Personally, I think this is an awesome step.

IX. Imagery

For those of you who prefer to learn through mental imagery, let's begin to wrap up this discussion by painting a picture of what we've just described. When envisioning the representation of the passage of time, instead of a time**line**, imagine a huge ring (a time**circle**). The ring has twelve sections—one for each month. Imagine that the ring is turning, rotating as the year goes on. Each section of the ring is one-twelfth of the total, and each section is rigged with a spotlight of a very particular color. Every month, planet Earth passes under the spotlight shining from one of the sections. This causes a unique influence because the color light that shines from that section is unlike the shade that shines from any of the other eleven sections.

We also need to understand that every year, a new ring begins. The ring for the year 5778 is distinct from the ring of 5779. This is because in 5779 Hashem created a new year of days that are pieces of the grand puzzle that are altogether separate from the puzzle pieces of 5778. However, there are many similarities between the two rings. That is, the colors and shades that shine from the Nissan section in 5778 are the same colors and shades that shine from the Nissan section in 5779.

This "5779 ring" is directly beneath the "5778 ring" and rotates at the exact same rate as the one above it. Hence, when planet Earth moves under the spotlight of the Nissan section, it receives the same shade of light that it received the previous year and the year before that. The only exception would be if the actions of the Jewish people in 5778 caused an impact on the month of Nissan. In such a case, when planet Earth passes beneath the Nissan section in the year 5779, the original shades of light that shone during the time of all the previous rings will be joined by the shades added in the ring of 5778.

Is everything clear? We could really start the *sefer* at this point, but for the sake of completion, let's add three more steps to this introduction. Feel free to skip the next few sections. They are confusing and it is not essential to understand them in order to learn this *sefer.*

X. Month Rings

Heretofore, we have focused our discussion on the months. Therefore, the rings described could be called "year rings"—in which each of the twelve sections is one-twelfth of the whole. However, to provide a complete picture of the reality, we should add that there are other rings that circle above our planet. For example, there are the *Yom Hachodesh* rings. This means that there are rings—the same circumference as the twelve-part "year rings" described above—but instead of twelve relatively large sub-sections that encompass a long period of time (i.e. a year), these "month rings" have thirty subsections and complete their rotation every thirty days. Being that these rings are divided into thirty sections (one for each day of the month), Earth passes under a new section every twenty-four hours. Each section has its own unique colors and illuminations. That would mean that on the thirteenth of every month, a particular light—which doesn't shine on any other day of the month—shines forth. To understand this in the context of the year rings, we can say that on the thirteenth of Tammuz the influences on that day will be from a mixture of the spotlight from the Tammuz section of the year ring and from the thirteenth section of the *Yom Hachodesh* ring.

XI. Week Rings

Then there is another set of rings called the "week rings." Remember how earlier we described how each day of the week has its own qualities? Based on that understanding, we must explain that each week that Hashem created has its own ring. It has the same circumference as the twelve-part "year rings" and the thirty-part "month rings" discussed above, but instead of twelve or thirty subsections, each "week ring" has seven subsections. These "week rings" rotate at a much faster pace than the "year rings" or "month rings." Every seven days, a new "week ring" begins, while a new "month ring" begins every thirty days and a new "year ring" only begins once a year.

We explained earlier that every time Earth passes under the Nissan section of the year ring, the Nissan colors and shades shine on Earth, and that every time Earth passes under the fifteenth section of the month ring, the colors of the fifteenth shine. Similarly, we can explain the influence of the week rings in this fashion. Every time Earth passes under the Sunday section of the week ring, the particular and unique shade of Sunday shines on the Earth. On Monday, however, a different shade shines onto the Earth. Every Monday receives the same Monday-type shade of light, in spite of the fact that every Monday is a brand-new piece of the grand puzzle.

This would mean, for example, that every Sunday in Nissan receives the influence of the Nissan-Sunday lights, while every Monday in Nissan there is a Nissan-Monday light shining. Monday in Iyar is somewhat similar to the latter because it has the Monday shade of light, but it's an *Iyar*-Monday shade, not a *Nissan*-Monday shade. Additionally, we can say that if the thirteenth of Tammuz falls out on a Monday, the influences on that day will be from a mixture of the spotlight from the Tammuz section of the year ring, from the thirteenth section of the *Yom Hachodesh* ring, and from the Monday section of the week ring.

XII. *Yom Haparshah* Rings

To sum it up, let's throw in information about one last ring—the *Yom Haparshah* ring. You see, every week we read a different portion of the Torah until, at the end of the year, we complete the entire Torah scroll. The *parshah* that is read each week exerts great influence over the week in a way distinct from the other rings. This ring looks very similar to the "week rings" but, in reality, is on its own track.

To put this idea in context of the other rings, we can say as follows. If the ninth of Tammuz falls out on a Friday during the week that we read *Parshas Chukas*, there will be four different spotlights shining on the world on that day. First, there will be the influence of the Tammuz section of the year rings. Secondly, there will be the influence of the ninth section of the *Yom Hachodesh* ring. Thirdly, there will be the influence of the Friday section of the week rings. And lastly, there will be the influence of the "Friday of *Parshas Chukas*" section of the

Yom Haparshah rings. Theoretically, if something would happen on this day, it would require otherworldly knowledge to pinpoint which ring's influence caused the event to occur (see *Magen Avraham, Orach Chaim* 580:9).

Nissan

Following the Leader

We know that each month is unique and has a specific *avodah* (job) related to it. What is unique about Nissan? It's obvious that Nissan is a special month because it contains the holiday marking the most open display of Hashem's Hand that the world has ever seen. Usually, Hashem hides behind the laws of nature. But one time in history Hashem went all out, displaying His complete and total dominion over the entire world.

The *Bnei Yissaschar* (*Ma'amar L'Chodesh Nissan* 1:7) explains that the name of the month even alludes to this. The word "*Nissan*" is cognate with the Hebrew word for miracle—*nes*. It seems that the singularity of the month is that it is a month of miracles. This idea is not only true with regard to the past. The Talmud (*Rosh Hashanah* 11a) states: *In Nissan we were redeemed (from Egypt), and in Nissan we will be redeemed (from the final exile).* Add this to the teaching of our Sages that the final Redemption will be of a most miraculous nature, even dwarfing the redemption from Egypt. What then is it about Nissan that lends itself to miracles? Why would Hashem contravene the laws of nature that He Himself set up? If we can answer this question then maybe we can discover our *avodah* for this month.

Let's take our first step in this investigation by analyzing the *mazal* (constellation) of the month—*Taleh* or Aries, the Male Sheep. What is the connection between sheep and Nissan? The first connection that comes to mind is that of the *Korban Pesach*—the special sacrifice brought on Pesach that was also a sheep. Immediately prior to the redemption from Egypt, we were commanded to take the sheep-god of the Egyptians and slaughter it in front of them. This courageous act would provide the great merit we needed to be redeemed.

This explanation does not suffice for two reasons: First of all, why should the slaughter of the Egyptian god be emplaced for all generations as the singularity of the month? The pagan Egyptian Empire isn't around anymore. Granted, it may have been important to debunk their deity at the time of the Exodus. But why is the slaughter of the sheep significant in modern times? Secondly, the *Bnei Yissaschar* writes (in numerous places) that each month's quality can be used for bad and good. Therefore, it is illogical that the unique quality of the month is something solely negative. What is the positive aspect of sheep?

Let's break it down. What do sheep signify? Everyone knows that sheep are the consummate followers. They are obedient and submissive. Is this a good quality or a bad quality? I suppose it depends. One can heed his friend's command to go down the worst of paths or up to the greatest of heights. It is not intrinsically negative to follow. It just depends on whom you are following.

Let's take this a step further. Every man is a follower. Some people follow others. But even the independent man follows—himself. He follows the dictates of his own mind and heart. We are taught that Egypt was a decadent country. They were filled with immorality, egoism, and all sorts of depravity. If the Sages teach that the Jewish nation sank to the forty-ninth level of impurity during their tenure in Egypt, we can only imagine how low the Egyptians themselves were! This essentially means that they were sheep—followers who gave in to the dictates of their basest desires. The Mitzri can be summed up in the personification of the sheep that he worships because he too is a submissive follower—of the lowly animalistic part of himself.

But it's not altogether wrong to be a sheep. Did you ever notice that immediately upon being granted our freedom from Egypt, the Jewish people accepted upon themselves the yoke of a new Master—Hashem? *For He is our G-d and we are…the sheep in His charge* (*Tehillim* 95:7). Why would a free nation willingly submit to a new Master, especially one as demanding as Hashem? The answer is that we recognized the great benefits of being the sheep of such a loving Shepherd. Would you not agree to obediently follow someone who wants to bring you up to the highest peak?

Now we can understand why the *mazal* of Nissan is a sheep.

Nissan is the month in which Yisrael was separated and chosen as a distinct nation from the other nations. But the quality that will set us apart depends on what type of sheep we choose to be. Will we be like the Egyptian sheep and obey the whims of our baser instincts, or will we fight our natural, selfish tendencies in order to serve the Higher Power and be sheep of the Almighty? It's not easy to choose the latter. The Vilna Gaon writes that the covenant at Har Sinai consisted of each party cutting something from themselves to give to the other party. Hashem, as it were, gave of Himself by giving His Torah, and the Jews gave of themselves by giving up the natural human desire to be independent.

Now we can suggest an answer to our original question. What is it about Nissan that lends itself to miracles? The answer can be found in the Netziv (*Ha'amek Davar, Yisro* 19:6): *If [the Jewish People] act toward Heaven merely within their nature, then [Hashem] will watch them with a Providence that is within nature. But if [Yisrael] serves Hashem by going above and beyond their nature, then Hashem will respond with a Providence similarly above nature.*

What this means is that Hashem willingly contravenes nature for us, but only if we similarly contravene our natures for Him. The redemption from Egypt was only possible if the self-serving, egotistical sheep was slaughtered. We would be worthy of miracles if only we would be courageous enough to stand up to ourselves to do what's right, not just what we feel like doing. This is also a way we can bring the final Redemption. Hashem is ready and able to bring those supernatural miracles that will free us from this long exile. But the way we can inspire Him to do so is if we too go against our self-centered natures to serve Him.

Yes, the task is difficult. The human tendency is to resist any attempts that others make to control us. We naturally want to be independent. Let's use Nissan as the impetus to become Hashem's sheep, and in that merit, may we indeed see the final Redemption.

From Slaves to Supermen

The month of Nissan is definitely a busy (and expensive!) month. We are usually occupied during these days with all sorts of important things, but let's not forget about growing in the unique Nissan *avodah*. What is the special quality of the month of Nissan? Which mountain must we climb during this month?

Let's take the first step in our journey by looking at the *Bnei Yissaschar.* The *Bnei Yissaschar* (*Ma'amar L'Chodesh Nissan* 1:7) focuses on the etymology of the name of the month. He explains that Nissan comes from the Hebrew word *nes*—miracle. Seemingly, this should be very easy to understand, for there is no month in the year during which greater miracles happened than Nissan. As a matter of fact, the Ramban (*Shemos* 13:16) writes that the supernatural display of miracles during the Exodus remains the one single time in history that Hashem revealed His Hand fully and completely—thus obviating the need to ever do so again.

But while it is true that Nissan *was* the month of miracles in the past, what is the significance of the month's supernatural quality for the modern-day Jew? Even if we consider the Gemara in *Rosh Hashanah* (11a) that states that the miracles of the future Redemption will also take place in Nissan, we still would need to discover the *avodah* of Nissan for all of us Jews sandwiched between the original and future redemptions.

Since it is obvious that the unique mitzvos of the month reflect the unique character of the month, let's turn to one of the two Biblical mitzvos that we observe during Nissan—*achilas matzah* (eating matzah at the Seder). Why do we eat matzah on Pesach? The Haggadah seems to contradict itself regarding this matter. At the beginning of the Seder we lift the matzah and declare, "This is the poor man's bread that we ate in Egypt." Seemingly, matzah is a symbol of slavery. On the other hand, in the middle of the Seder we state, "This matzah that we eat is because our dough had no time to rise, because of the speed with which we were chased out of Egypt..." Additionally, the halachah is that we recline (to display freedom) when we eat matzah, as opposed to when we eat *maror*, the bitter herbs. Apparently, matzah is able to simultaneously represent slavery and freedom. If we can explain this duality, we will be able to discover the *avodah* of Nissan.

Let's turn to the second Biblical mitzvah of the month—*sippur Yetzias Mitzrayim* (retelling the story of the Exodus at the Seder). The Gemara (*Pesachim* 116a) describes how this mitzvah is properly fulfilled: *You start [by describing] the lowly parts, and [after that, you describe] the praiseworthy parts*. This means that we first describe how terrible it was in Egypt, and afterwards we describe the great salvation Hashem performed for us. The Gemara continues: *How do we fulfill this requirement? Rav says that we say, "Originally (in Egypt) we were* ovdei avodah zarah. *[But Hashem redeemed us and brought us close to His Service.]" Shmuel argues and says that we recite* Avadim Hayinu. *"[Originally] we were slaves [to Pharaoh in Egypt. But Hashem redeemed us from there…]."*

If you look in the Haggadah, you will see that we fulfill both opinions. The Vilna Gaon (*Gra al Haggadah*, s.v. *mitechilah*) explains that we do both because they are both true. He writes: *There were two exiles to which we were subject—an exile of the body and an exile of the soul. [We therefore thank Hashem for redeeming us from both.]*

What does this all mean? We are all familiar with the idea that Shmuel mentioned about a physical exile. But what does Rav mean when he refers to a spiritual exile?

To answer this question, let's conduct a technical analysis of the phrase we just mentioned: *Originally (in Egypt) we were* ovdei avodah zarah. What does the Hebrew word *ovdei* mean? Most people would say it means "worshippers." The sentence would then read: *Originally (in Egypt) we were worshippers of idols*. The problem with this definition is that we find the same Hebrew root of *ayin beis dalet (**EVD**)* in the phrase ***Avad**im hayinu l'Pharaoh b'Mitzrayim*. In that context, the root ***EVD*** clearly does not mean worshippers—for then it would read, *We were originally **worshippers** of Pharaoh in Egypt*. Rather, that context teaches us that ***EVD*** means "slaves," as in, *We were originally **slaves** to Pharaoh in Egypt*. This definition of ***EVD*** makes it a little more difficult to understand Rav, because if ***EVD*** means "slave," then the phrase, *We were originally* **ovd**ei avodah zarah, means that we were originally **slaves** to *avodah zarah.* A slave is one whose physical actions are constrained and dictated by a master. An idolater, on the other hand, is someone who (literally) idolizes his deity and ostensibly fulfills his commands out of free will. So why does Rav call idolaters "slaves"?

The answer lies in a deeper understanding of what makes people worship idols (or commit any sin, for that matter). Why would a holy soul decide to worship foreign gods? The answer is—the *yetzer hara,* the evil inclination. While sometimes we feel like serving Hashem and fulfilling His Torah, we all too often feel a compulsion to defy His Will. It's not that we are schizophrenic; it's just that the human being is made up of two polar opposites—the animalistic body and the spiritual soul. Life is a constant struggle between these two forces.

The goal of every good Jew is to strengthen the soul, day by day, until it is master over the body and its desires. However, what would happen if, G-d forbid, the body would gain supremacy? Essentially, the soul would be forced to concede to the body's every whim. In other words, we would become slaves—to our own physical desires. We stated earlier that a slave is one whose physical actions are dictated to him. Is there any greater example of a slave than one who is forced to act on his every desire because he cannot control himself? The reason Rav called an idolater a slave is because one who cannot control his inclinations and desires can indeed be labeled a slave—to his own body.

Armed with this understanding, let's backtrack and explain the duality of matzah. In reality, matzah is a poor man's bread. Yet it is also a symbol of freedom because *a truly free man can choose what he wants to eat.* He can eat rich bread or even a poor man's bread. He is the master of himself and can decide in what he should indulge, in what amounts, and when. A slave, on the other hand, is forced by his body to indulge—even against the better judgment of the soul. So, while it's true that in Egypt we suffered from an exile of the body, let's not forget that we also suffered from an exile of the soul. We were ***ovd**ei avodah zarah*—slaves under the dominion of our inclinations and bodies. Hashem redeemed us from this exile by raising us up to become His holy nation. By the time the Exodus was complete, we were no longer forced to obey the body's compulsions. We were free.

Now let's return to our original question about what exactly is the *avodah* of Nissan. Perhaps the answer is that Nissan is the time to focus on the supernatural powers granted to us by the Torah. By fulfilling the Torah and mitzvos we can increase the power of the soul and enable it to dominate the body. We can use the Torah to rise above the natural desires and tendencies with which we were born. Instead of living by the liberal philosophy that each man is a slave to his predispositions, Judaism allows us to conquer our bodies and become supermen. All we need to do is take a few steps on our own and Hashem will surely bless us with added help from Above, just as He did in times of yore.

Talk the Talk vs. Walk the Walk

Every year, when Nissan arrives, Jews around the world exclaim, "I can't believe it's already almost Pesach!" Why does Pesach take everyone by surprise? Either way, we should definitely also be asking ourselves, "What is our *avodah* for the duration of this month? How does Hashem want us to grow while we are on this page of the calendar?"

The *Sefer Yetzirah* writes that in Nissan we are enjoined to rectify *sichah*—speech. How is this to be understood? What must we do to speak properly? And why does Pesach fall out in the month of speech?

Perhaps we can discover the answer to these questions by embarking on an analysis of the mitzvah of Pesach. You're probably wondering—which one? After all, are there not many mitzvos of Pesach? The answer is that yes, there are. We must eat matzah and *maror* (bitter herbs), we drink four cups of wine, we sing Hallel, etc. But it's important not to lose sight of the fact that all these mitzvos relate to the one central mitzvah of Pesach: *sippur Yetzias Mitzrayim*—recounting how Hashem took us out of Egypt. *Maror* recalls the bitter slavery. Matzah recalls the speedy redemption. The four cups recall the four aspects of our redemption—how Hashem took us out, saved us, redeemed us, and took us to Him for a nation. Hallel praises Hashem for fulfilling His promise. All these mitzvos are important, but we must recognize that they all revolve around the proper fulfillment of one mitzvah—*sippur Yetzias Mitzrayim*.

I believe there is a nuance in the phraseology of the mitzvah that bears noting. The mitzvah is called ***sippur*** *Yetzias Mitzrayim*—***telling*** the story of how Hashem took us out of Egypt. Why is there an emphasis on the telling? Isn't the main idea to *remember* how Hashem took us out? Shouldn't the mitzvah have been termed ***zechiras*** *Yetzias Mitzrayim*—***remembering*** how Hashem took us out?

The answer lies in understanding that *telling over* the story is part and parcel of the mitzvah. It wouldn't be sufficient to merely remember the story. Even if everyone at the table already knows the story, it still must be told. One person tells it to his friend and then his friend tells it to him. Even if two adults are alone at a Seder, they are halachically required to ask each other the *Mah Nishtanah*—the Four Questions. One may not simply eat the matzah and *maror* and *remember* how the slavery was bitter and how Hashem redeemed us speedily. One must *tell over* how the slavery was bitter and how Hashem redeemed us speedily. That is why the mitzvah is not called *zechiras Yetzias Mitzrayim*, but rather *sippur Yetzias Mitzrayim*.

However, this needs explanation. Why must the story be told over? Why is it insufficient to recall Hashem's salvation? Perhaps we can find the answer if we take a detour to an event that took place on Pesach hundreds of years prior to *Yetzias Mitzrayim*. Let's travel back to the battle where Yaakov and Eisav fought over who would be the progenitor of G-d's Chosen Nation. Of course, we know that Eisav abdicated his rights when he sold his firstborn status for a bowl of soup. But on that fateful Pesach in the year 2171, Yaakov was forced to deceive his father in order to receive the blessings from which Eisav was disqualified due to the sale and his wickedness (neither of which Yitzchak had any knowledge).

Yaakov dressed himself in goat skins so if his blind father would hug him, he would feel like his hairy brother. He did not worry that his voice would give him away, because he and his twin sounded alike (*Rashbam*). Yet when he entered with the requested Pesach sacrifices and spoke with his father, Yitzchak grew suspicious as to whether his visitor was really Eisav—the son whom he was expecting. After feeling Yaakov's hairy arms, Yitzchak declared, "The voice is the voice of Yaakov, but the hands are the hands of Eisav!" Now, as we mentioned earlier, Yaakov and Eisav sounded very alike. So what was it that made Yitzchak say that this visitor sounded like Yaakov? Rashi answers that Yitzchak was suspicious because of two things that Yaakov said. First, when Yaakov entered, he spoke to Yitzchak in a gentle and soft tone. He said, "*Please* sit up and eat of my game." Eisav did not speak that way. When Eisav subsequently came in with his game, he ordered, "Get up, Father!" The second thing that tipped Yitzchak off was that when he asked, "How did you get here so quickly?" Yaakov answered, "Hashem arranged it for me." Eisav was not one who would regularly mention Hashem in his conversations. These two things made Yitzchak suspicious that this visitor was Yaakov and not Eisav.

We see from here two lessons in the way a Jew speaks. First of all, a Jew speaks softly. He requests; he doesn't demand. But the second thing we see, which I think is more relevant to our discussion, is that a Jew incorporates Hashem into his speech. Or, more precisely, he gives credit for everything he has to the One Who gave it to him. Yaakov didn't answer that he got lucky and was able to trap two animals quickly; he rather said that Hashem arranged it for him. This is part of the essence of Yaakov—and that's why Yitzchak immediately picked up on it. Eisav didn't speak like that. Eisav would have taken the credit for himself.

Now we can explain why Pesach is about *sippur*. Remembering the story is important, but on the birthday of the Jewish nation we work to inculcate this supremely Jewish trait of *articulating* our recognition of Hashem's Hand and verbally giving credit where credit is due.

By now we should be able to suggest how to rectify *sichah* and why it's specifically done in Nissan. In Nissan, we focus on learning to talk like Jews.[1] There is such a big difference between "I found a parking spot" and "Hashem gave me a parking spot." It may seem inconsequential, but it can change your life if you make it a habit. We say "*baruch Hashem*" and "*im yirtzeh Hashem*," but we often forget to think about what these phrases mean. I would suggest we say them in English, so we start meaning what we say. Instead of "I'll see you tomorrow," say "With Hashem's help, I'll see you tomorrow." Instead of saying "That was a good supper," say "Hashem gave me a good supper."

Let's learn to speak like *Yidden.*

Freedom to Be...

Pesach is by far the most detailed holiday in our calendar. Serious students hit the books to plumb the infinite depths of the Haggadah in an attempt to pull out new understanding and insights. Indeed, new commentaries on the Haggadah are printed each year; it seems that there is no limit to how much we can learn from the story of Pesach. Then again, a common downside of having so much material is that it is easy to get overwhelmed. There is so much to learn from Pesach; where do we start? Let's attempt to pull out a core lesson of Pesach and see if, without oversimplifying, we can discover what Pesach is really all about.

---

1. See *Pri Tzaddik, Ma'amar L'Chodesh Nissan, os* 2: *[In Nissan,]...the reigning faculty is* sichah*...wherein we open our mouths to praise Hashem...like the well-known* remez (hint) *that Pesach should be read* peh-sach*—the mouth that speaks... and the fact that the Jewish people sang* shirah (song) *also ties in with this theme that Nissan is about* sichah...

The Haggadah (ostensibly) centers around the story of the Exodus. Indeed, the Haggadah says that when our child asks, "Why is this night different from all other nights?" we answer, "*Avadim hayinu l'Pharaoh b'Mitzrayim*—we were slaves to Pharaoh in Egypt. *Vayotzi'einu Hashem misham...*—and G-d took us out from there..." Apparently the Seder night is special because it is the night G-d freed us from Egypt. One would therefore probably be correct in assuming that the crux of the holiday is focused on freedom.

Freedom! What a wonderful word! We will leave the evil master who would dare tell us how to live our lives! We are free to do whatever we want!

"Freedom" is a very popular word in America. We have a Bill of Rights to protect our Freedom of Speech, Freedom of Religion, Freedom of Press, etc. That's why even assimilated American Jews understand Passover. After all, freedom is *the* core value on which the United States is based. The national anthem itself calls America "the land of the free." So have we discovered what Pesach is all about? Is Pesach the holiday that celebrates our freedom, our ability to be our own bosses and do what we want? Let us study the story of the Exodus a little more, and then we'll return to our question.

Immediately following the Exodus from Egypt, we began the forty-nine-day trek to Har Sinai. Once there, Hashem appeared to all of us in a revelation unmatched to this day. Amidst thunder and lightning, G-d gave us the Torah. What is the Torah? Overall, it seems to be a rulebook—and not just any rulebook, but a demanding one too! The Torah encompasses every aspect of our lives. It dictates how to eat, how to sleep, how to walk, how to talk, how to think, and even how to feel! The Torah dictates what we're permitted to wear, what we're permitted to say, and even what we are permitted to look at! It would almost seem that in exchange for one master, we accepted upon ourselves a new Master Who was much more demanding. Is this freedom?

It is said that there are two ways to tackle a question. One way is to acknowledge the question and then answer it. The other way is to challenge the assumption upon which the question is based. In this case, I would like to attempt the latter.

The assumptions upon which our confusion is based are:

1. We are assuming that our understanding of freedom is correct, that freedom is, as Webster's Dictionary puts it, "the power to act or speak or think without externally imposed restraints."
2. We are assuming that such freedom is a good thing, an end that we would strive for, and an end that a benevolent G-d would bestow upon us.

Now that we have specified our assumptions, let's dispute them.

Here's step one: Let's ask ourselves, why does G-d give us all these commandments? Why does it matter what we eat, what we look at, and what we think? Are these commandments random, given by a capricious god? Does Hashem really care what I say or do? Isn't Hashem perfect and omnipotent? What does He gain by having me do all this?

The answer is that Hashem doesn't gain anything when we fulfill His commandments. Our performance of the mitzvos does not affect Hashem whatsoever. He is above that. So why does He command us to perform them? The answer is, for *our* benefit. The only reason why Hashem cares about what I do is because He wants *me* to become a better person.

Hashem wants us to live holy lives. Hashem wants us to become like Him because, ultimately, that is the greatest pleasure possible. Just as a caring doctor will provide a patient with an extremely detailed diet, the Doctor of all doctors provides us with a health plan that can make us into G-dly people, indeed a "light unto the nations." Each and every one of the mitzvos makes us more holy. Each mitzvah brings us closer to G-d.

This understanding takes us one step closer to what we need to know to address our earlier points. However, we're not fully there yet. Let's take another step.

Let's picture the following experiment: What do you suppose would happen if a nation was granted its "freedom"? What would the long-term effects be if a group of people were given "the power to act or speak or think without externally imposed restraints"? It's probably not too hard to imagine the results. In the name of Freedom of Speech, one would be allowed to lambast any individual or group with impunity. In the name of Freedom of Press, one would be protected by law (and perhaps even paid!) to find juicy, embarrassing stories about people and publish them for the public's entertainment. Abominations would be legalized, deadly weapons would be available to the public, and perverse reading and viewing material would be widely disseminated—all in the name of "freedom." Society would degenerate to such a point that anything could be defended and legalized under the banner of "rights." It is hard to imagine a nation so conceived and so dedicated to this god of freedom being able to long endure.

Now let us contrast (in this theoretical experiment) to what society might look like if it were run *with* external (G-d-imposed) restraints. Speech would be restricted, in that derogatory speech of any kind would be absolutely forbidden. Reading or viewing materials that contain ideas contrary to holiness would be banned. Abominations would not be tolerated. Overall, the mindset of society would be one of growth and commitment to a proper way of life. Imagine such a utopia—where nobody is out to get you and everyone only wants to help each other. Sounds pretty good, no? With these thoughts in mind, let's return to our assumptions. I think we may see them a little differently now.

Mr. Webster had said that freedom is "the power to act or speak or think without externally imposed restraints." Do we really want a society without any externally imposed restraints? As we demonstrated, such a society would surely be corrupt. One cannot hope to have a moral nation if there is no higher moral code to which said nation adheres. **If morality is subjective, it has no meaning.** In addition, no benevolent god would condemn his people to a life without guidance. Hashem, in His infinite love for us, gave us the Torah. The word "*Torah*" means "Instructions." G-d instructed us in His ways so we could truly be His treasured nation and a holy nation. Hashem wanted us to have a perfect society, so He (lovingly) imposed external restraints on us.

So now we know what Pesach is *not*. But do we know what Pesach *is*?

I think the answer can be encapsulated in a quote from the Gemara in *Megillah* (14a): *Praise the servants of G-d, who are not the servants of Pharaoh.* This means that we *are* still servants. But now we are the servants of G-d. What a wondrous position to be in! We are privileged to have been chosen to serve the almighty Creator of the heavens and the earth! Moshe, the greatest of our leaders and greatest prophet of all time, was granted the title of *Eved Hashem*—Servant of G-d. We see from here that there is no greater title than Servant of the King.

Now perhaps we can understand the purpose of the Exodus. Yes, G-d freed us from Egypt. Yes, G-d removed the yoke of our affliction and granted us the "power to act or speak or think without externally imposed restraints." But for what purpose? Freedom is not an end unto itself. G-d only granted us our freedom so we'd be free to serve Him. Hashem removed the *externally imposed restraints* that Pharaoh placed on us so that we would be able to pursue a higher calling—that of a life of meaning and holiness.

I think this idea is implicit in the Haggadah. We find that the Haggadah describes how Hashem used four "phrases of redemption" (*Shemos* 6:6-8):

1. *V'hotzeisi*—and I will take you out.
2. *V'hitzalti*—and I will save you.
3. *V'ga'alti*—and I will redeem you.
4. *V'lakachti eschem Li l'am*—and I will take you to Me for a nation.

Is it not obvious that the redemption was a package deal? Hashem didn't want to just *take us out, save us, and redeem us* so that we could wander around in the wilderness eating *mann* all day. Hashem only wanted to *take us out, save us, and redeem us*, so that He could *take us to Him for a nation.*

We also find that the Ramban expresses this idea in his introduction to *Sefer Shemos*: *The exile did not end until the day the people...returned to the stature of their forefathers. When they left Egypt, although they had left the "house of bondage," they were still considered exiles. But when they arrived at Mount Sinai and built the Mishkan and the Holy One, blessed is He, once again caused His Divine Presence to dwell among them...then the People of Israel were truly considered redeemed from their exile.*

This, my friends, is what we celebrate on Pesach. We rejoice in the fact that Hashem freed us from slavery and chose us as His nation. This is the freedom that we desire. Any other freedom would be meaningless and even detrimental. Praised be G-d that we no longer have to serve Pharaoh. And all the more should G-d be praised now that we are able to serve Him instead. Pesach is about freedom—freedom to be servants of the Almighty.

Pesach: Relationships and Sandwiches

Nissan is perhaps the busiest month of the year. A tremendous amount of work is done, all in preparation for Pesach. From the moment Purim ends (and even earlier for some people) all people talk about is Pesach, Pesach, and Pesach. Now, if this word is to resound in our heads for so long, we should at least understand what it means. After all, the Torah teaches that a name denotes a thing's essence. So let's ask a very basic question: What is the meaning of the name "Pesach"?

We are probably all aware that the word *Pesach* literally translates as "Passover"—a term used in the verse (*Shemos* 12:27) that states that, *Hashem **passed over** the houses of the Children of Israel when He smote [the firstborns of] Egypt.* This verse seems to indicate that the holiday's name came from the last of the Ten Plagues. This seems strange. Why would this particular detail be chosen to sum up the essence of this holiday? Even if you will argue that it was only after this plague that Pharaoh freed us, one can still wonder—was Hashem's passing over our houses really the most important aspect of the story of our Exodus?

Interestingly, it's not only the holiday that carries the name Pesach. In the days of the Beis Hamikdash, we brought a *korban* (offering) on this Yom Tov which was also called a *Pesach*. While one might think that the *Korban Pesach* was called thus because it was offered on the holiday of Pesach, a careful reading of the relevant verses dispels such a notion. The Torah reads (*Shemos* 12:27): *And they said, "It is a Pesach offering to Hashem, Who* ***passed over*** *the houses of the Children of Israel when He smote [the firstborns of] Egypt."* From here we see that both the holiday and the offering were independently called "Passover," and both were to commemorate Hashem skipping over our houses in Mitzrayim. Why is this detail so important?

Let's add a second question about the *Korban Pesach* to the mix. The Haggadah states: *Rabban Gamliel used to say, "Whoever doesn't say these three things on Pesach hasn't fulfilled his obligation: Pesach, matzah, and* maror*."* In this case, "*Pesach*" is referring to the offering, as evidenced by the subsequent elucidation that begins with the words, *This Pesach that we would eat…* This teaching of Rabban Gamliel is hard to understand. *Maror* and matzah are obviously central themes of the holiday. *Maror* represents the bitter enslavement and matzah represents the under-eighteen-minutes redemption. But where does the *Korban Pesach* fit in? What makes it so important that mentioning it is crucial to the fulfillment of the Seder?

Before we answer these two questions, let's add one more. We know that one of the mitzvos at the Seder is called *Korech*. This is when we take the *Korban Pesach*, matzah, and *maror* and eat them all together. What is the purpose of this? All three of these mitzvos are fulfilled individually at different parts of the Seder, so why must we do them all again? What is gained when we eat these three things as a sandwich?

In order to answer all these questions, let's take a detour to analyze the laws of the *Korban Pesach*. This offering has an inordinate number of mitzvos attached to it. The sheep must be roasted and not cooked in water. It may not be cut up and must be put on the spit whole. The Torah forbids breaking a single bone in the sheep's body. The list goes on and on. Why does the *Korban Pesach* have these particular details? To make matters even more strange, the *Korban Pesach* is called ***ha***_'avodah_—the service [of Hashem], almost implying that this offering is the prototype for all others. Yet no other *korban* is like this one! How can we understand this?

Addressing all of these questions, the Maharal explains that all the laws of the *Korban Pesach* are meant to direct us to internalize one idea—namely, the oneness of Hashem. The world is not controlled by good forces and bad forces that struggle for supremacy. Rather, everything comes from one Almighty Creator. To allude to this idea, the *korban* must be kept in one piece. The bones must remain whole and may not be broken in two. It cannot be cooked in *mayim* (water)—which is a word that only exists in the plural; rather, it must be roasted—a method that removes all *mayim* and compacts the meat into one piece.

So let's return to our initial questions. It has been said that if Judaism had to be summed up into a single concept, it would be that of a relationship with Hashem. The entire Torah is directed toward this goal. This relationship began (on a national scale) on Pesach. Pesach is the birthday of the Jewish people. Until the Exodus, we were a disparate group of people descended from common ancestors. But the moment Hashem *passed over* our houses and struck only the Egyptians was a very historic one. It was at that second that Hashem segregated us from the nations. We were set apart and made into a unique nation—the Children of Hashem. The fact that Hashem passed over our houses during the final plague is not a minute detail. Hashem choosing us as His nation is the sum total of the holiday. It was only through the passing over that our relationship with Hashem began.

However, as the expression goes, it takes two to tango. By definition, a relationship is comprised of two parties. Hashem did His part; the Jews still had to do theirs. The job of the Jewish people is to do *avodah*—service of Hashem. *Avodas Hashem* is the tool that we use to form a relationship with the

Almighty. Every mitzvah brings us closer to Him, and every *aveirah* (sin) distances us from Him. When the predestined time for the Exodus came, the Jews were lacking any merits to be freed. They had no mitzvos that made them unique as a people or worthy of being passed over when Hashem punished the Egyptians. So Hashem provided the *Korban Pesach*, which was the prototype mitzvah in that it tested the new nation to see if they would take that step to initiate a relationship with the Creator. Performing this service, with all its requisite laws, would show that the people believed in the one and only Master of the heavens and the earth. By offering up this *korban* they would be declaring their willingness to begin working on a lifelong relationship with G-d. Then, and only then, would they be worthy of being passed over. Only through *avodah* are we Hashem's people.

Now we can understand the *Korech* sandwich. This sandwich is made up of three ingredients—the *Korban Pesach*, matzah, and *maror*. As mentioned, the *Korban Pesach* declares the oneness of Hashem. This idea of oneness is essential if we are to have any relationship with the Creator. He is omnipotent, and *everything* comes from Him. Nothing happens in the world if He doesn't decree it first. All good things that happen to us (read: matzah) are coming from His Will, as are all negative things that occur (read: *maror*). We are very quick to blame G-d when bitter things happen; with happy occasions, less so. But our *avodah* is to wrap it all together and "ingest" the idea that there is but one Source for everything that occurs. If we can realize that it's the same all-powerful and loving Father Who sends the good and the bad, perhaps we'll merit seeing how the bad is, in reality, good.

Matzah: The Message of Might

It's important not to go into Pesach without knowing more than we did when we packed away the boxes last year. To gain a greater appreciation of this central Yom Tov, let's focus on the mitzvah from which the holiday of *Chag Hamatzos* gets its name—matzah.

We all know that there is a mitzvah at the Seder (or, according to some, all throughout Pesach) to eat the unleavened bread called matzah. What is the reason for this commandment? The Haggadah cites the teaching of Rabban Gamliel: *This matzah that we eat is because our dough had no time to rise, because of the speed with which we were chased out of Egypt...* This mishnah tells us that of the many symbols at the Seder to recall the slavery and the redemption, the matzah fits into the latter category. It's a sign of redemption, just like the Four Cups and the tradition of leaning. On the other hand, the Haggadah also adjures us to lift the matzah and declare, "This is the poor man's bread that we ate in the land of Egypt." This excerpt seems to indicate that matzah is actually a symbol of slavery—just like the bitter *maror.*

What is the meaning of these conflicting messages? It seems odd that we would make separate remembrances throughout the Seder (i.e. *maror* vs. leaning) for the respective aspects of slavery and redemption, and then make one remembrance (i.e. matzah) for both slavery *and* redemption. Is it a coincidence that two opposite goals are accomplished with one item? Let's delve deeper into the matter and see if we can explain that it's not a coincidence after all.

Let's take our first step by analyzing a different mitzvah of the Seder: *sippur Yetzias Mitzrayim*—recounting the story of the Exodus. The Gemara (*Pesachim* 116a) describes how this mitzvah is properly fulfilled: *You start [by describing] the lowly parts, and [after that, you describe] the praiseworthy parts.* This means that we first describe how terrible it was in Egypt, and afterwards we tell of Hashem's great salvation. Indeed, there is much to say on this topic. The Egyptians were incredibly cruel to our people, and G-d did truly go all out when He redeemed us from their hands. But why are we supposed to mention both aspects—the lowly and the praiseworthy? I suppose the simple understanding is that recounting the lowly is not an end goal but rather a means through which we can truly appreciate the praiseworthy. If we don't know how bad it was, we won't be able to appreciate how good it became. However, I think there is room for a different answer. Let's see if we can find a way to explain that both aspects are the goal, not that one is the goal and the other an enabler.

To do this, let's ask a philosophical question. Why should we praise Hashem for taking us out of Egypt if He is the one who put us there in the first place? We would indeed be obliged to praise G-d if a power other than He cruelly enslaved us. But we know that Hashem is in complete control of the world, so why should we thank Him?

The answer is that we thank Hashem for both the sickness and the cure, because both are for our benefit. It's easy to see how a cure is beneficial, but Hashem also sends the sickness only because it's good for us! Sometimes it's an atonement for our sins. Sometimes it's a test that can increase our reward in the World to Come. Yet other times it helps form us into stronger people. Some people are shattered when difficulty comes their way, but Hashem expects us to become stronger in our faith and trust in Him.

I believe that this latter reason explains how we can praise Hashem for both taking us out of Egypt and putting us there in the first place. We gained not only from the redemption, but also from the slavery. From the all-out display of miracles Hashem made to save us, we gained a recognition of His absolute power. We saw His mastery over every aspect of the world. However, the fact that G-d is powerful might not necessarily preclude the possibility of us also having power. Perhaps G-d is powerful but we also can control our lives. Ah, but the absolute servitude to which we were subjected displayed beyond a shadow of a doubt that we are *not* in control of our destiny. True *bitachon* (trust in Hashem) has two parts—not only must we know that Hashem is all-powerful, we also must know that we ourselves do *not* have power. Neither is sufficient without the other.

Now I think we can suggest an explanation for why the mitzvah of *sippur Yetzias Mitzrayim* begins with describing the lowly and then the praiseworthy. They are both part of the goal. If we skipped the lowly parts then we'd only know of Hashem's might, but we may still entertain delusions of having some might of our own. But if we first describe our total subjugation, we come to realize that we are not in control. Once that option is off the table, *then* we can start discussing how Hashem is the One Who controls our lives.

Perhaps now we can backtrack and answer our original question. How can we understand the idea that matzah simultaneously represents slavery and redemption? The answer is that they're both the same idea. The slavery and the redemption *jointly* taught us the lesson of Hashem's absolute control to the exclusion of any other power. Hashem had a reason for putting us in Egypt in the first place. He wanted to refine us into His chosen nation, the nation that testifies to the truth about reality. We are the people who recognize the higher reality that *He is the one who grants strength* because it is ***not*** *the strength and might of my hand that has wrought this accomplishment* (*Devarim* 8:17).

This is the greatness of the Jewish people. We live with a relationship with the Almighty, recognizing that He is indeed all-mighty, while we are merely His servants. If we can tell the story of the Exodus and eat the matzah in a way that we will imbibe both aspects of this single lesson, we'll have accomplished the great message of the holiday of Pesach.

Iyar

Connectivity and Sensitivity

As we know, *Chazal* drop numerous hints to help us discover the specific area in which we must grow during each month. What is special about Iyar? The challenge is on; let's see if we can surmount it.

We know that each month is aligned with a particular tribe and letter. The commentators match Iyar up with the letter *vav* and the tribe of Yissachar. Both of these facts need explanation. The idea of the letter *vav* is simply obscure, and Yissachar doesn't seem to fit with Iyar either. This is because his tribe is well known as a tribe of scholars. This is seen clearly from Rashi in *Vayechi* (49:15) where he describes Yissachar as the one who *bends his shoulder to bear the yoke of Torah…to provide practical halachic decisions for Israel.* Shouldn't the tribe of Torah be aligned with Sivan—the month of *Mattan Torah*, when we received the Torah? Why is Yissachar matched up with Iyar—the month *before* we received the Torah?

Let's ask another question. During this time of year, the Torah requires us to count the days from Pesach leading up to Shavuos. We are all familiar with the idea of preparing for *Mattan Torah.* But how does counting the days help us become worthy of receiving the Torah? This question is even stronger when one realizes that if he misses a single day, halachah states that he may no longer continue counting (with a blessing). Why is there such a strict penalty attached to this mitzvah?

The truth is that the halachah *sefarim* address this point by explaining that all forty-nine "countings" are considered one long forty-nine-part commandment. Since each night's counting is only one forty-ninth of the mitzvah, if one misses one night, he can no longer fully complete the mitzvah. But why did Hashem set it up this way?

In order to answer these questions, let's take a detour and

analyze another aspect of Iyar—the restrictions of *Sefirah*. As we know, Rabbi Akiva had 24,000 students who perished in a plague between Pesach and Shavuos. The Sages subsequently enacted a series of restrictions (forbidding weddings, shaving, music, etc.) during this time period in order to encourage us to work on a particular area of improvement. Since Judaism doesn't believe in coincidences, it is eminently logical to conclude that the students died specifically during this time period for a particular reason. That is to say, not only must there have been a Divine reason for why they deserved to die, but the reason must also relate to this time of year. Ergo, when we discover their sin, we will also discover an essential *avodah* (job) specific to the month of Iyar.

The Talmud (*Yevamos* 62b) explains that Rabbi Akiva's students died because they didn't treat each other with the proper respect and honor. Now, these men were great Torah giants and their sin could not have possibly been any more than a minute infraction, discernible only by the Almighty. That being the case, one might very well wonder why Hashem viewed it as such a crime. Was it not enough that they studied Hashem's holy Torah day and night? What was so egregious about this mistake that caused these students to lose the right to carry the Torah over to the next generation?

The answer to this question can be found by gaining a more profound understanding of the Torah's purpose. The Torah is not only meant to be studied. The Torah must also be fulfilled. When Moshe went up to Heaven to get the Torah, the angels complained to G-d that they wanted to keep the Torah for themselves. Upon being commanded to respond, Moshe challenged the angels, "Do you have parents to honor or an evil inclination that incites you to covet?" We see from here that Hashem does not grant the Torah to anyone other than those who are able to fulfill it. A similar point is implicit in the *Sifri* (*Devarim* 343) that describes how Hashem offered the Torah to the nations of the world. When the nations responded that they did not want to fulfill it, Hashem did not respond, "So don't bother fulfilling it; just take it and study it!" Apparently, Hashem doesn't just want the Torah to be studied; He also wants it to be fulfilled.

This was the sin of Rabbi Akiva's students. We know that "*[the Torah's] ways are pleasant ways*" (*Mishlei* 3:17). If there was even a slight lack of interpersonal propriety on the students' part, that means that they were failing (in a minute way) to apply the Torah. Hashem expects a person to achieve his potential to the fullest extent of his capabilities. These holy men certainly applied what they learned to a great extent. But Hashem decided that they could have, and therefore should have, connected their studies to their lives in a greater way than they did. They may have been outstanding scholars, but what good was their erudition if they lacked the fundamental ability to (fully) connect what they learned to their lives?

Perhaps this can help us explain why Hashem wants us to count *every single day* from Pesach to Shavuos. If the Torah was merely one of the sciences, then it would be sufficient to include it in our lives on a semi-regular basis. But Torah is more than a science. The *Toras Chaim* is a Divine Instruction Manual for every day of our lives. Hashem wants us to grow in our ability to serve Him daily, so He instructed us to practice incorporating Divine service into our daily routines. No days off; if you skip a day, then you're missing the point. A Jew serves Hashem *every single day* of his life—without exception.

I think this can lead us to understand the lesson of the *vav.* According to Rav Hai Gaon, *vav* is a letter that connotes connection. When used as a prefix, the *vav hachibbur* (connecting *vav*) means "and" because it connects two distinct phrases. Additionally, the Hebrew word *vav* means "hook." A hook connects two disparate objects (see *Shemos* 27:10). The *avodah* of Iyar lies in learning how to apply what we know about Hashem's will to our lives. A Jew must understand that he not only must *learn* the Torah (which we will focus on after *Mattan Torah*), but he must also *connect* the Torah to his daily life. Judaism is not just for the shul or the study hall. It's for every aspect of every day.

Let's sum this up by explaining perhaps why Yissachar is aligned with Iyar. You see, Yissachar is not the only tribe of

Torah. After all, we also have Levi—whose primary job is to study and teach the Torah. But Yissachar has a unique responsibility. As the Rashi quoted earlier taught, Yissachar was the one who provided practical halachic decisions for the Jewish people. In other words, Yissachar was the one who enabled Klal Yisrael to *connect the Torah to everyday life*. They turned the Torah into a *Toras Chaim*. Perhaps the reason Yissachar is placed in the month before *Mattan Torah* is because his trait of understanding that the Torah is to fulfill is a prerequisite to being worthy of receiving the Torah.

May we merit to grow in our ability to feel that every day of our lives must be based on Torah, and in that way may we make Hashem proud.

The Beauty of the Beast

Every good Jew knows that he must always be growing closer to Hashem, one step at a time. In His kindness, Hashem gave us different months of the year to work on different qualities. What is the area in which we must grow during the month of Iyar? Is there something specific Hashem has in mind? Come—our journey has now begun.

According to the opinion of the Arizal, Iyar is aligned with the tribe of Yissachar. Yissachar is definitely a great role model, as he was a tremendous Torah scholar. The *pasuk* states about him: *Yissachar is a strong-boned donkey...he bent his shoulder to bear [the yoke of Torah—Rashi]* (*Bereishis* 49:14-15). However, notwithstanding Yissachar's greatness, one could ask a fair question. Shouldn't the Torah scholar be aligned with Sivan—the month of the Torah and Shavuos? The fact that Yissachar was placed in Iyar makes it apparent that it's not his erudition that we must imitate, but rather a different quality in which he excelled. What is that quality?

Before addressing this question, let's add another. We know that our Sages teach that not only does each month have a tribe, but it has a *mazal* (constellation) as well that serves as a model of how we should act. Iyar's *mazal* is the *Shor*—Taurus, the Bull. What is it that makes a bull special? In what way should we emulate it?

In order to answer these questions about Yissachar and Taurus (and thereby discover the *avodah*, specific job, of Iyar), we will need to take an extended detour to understand the one special mitzvah that applies every day of this month. Of course, we are referring to *Sefiras Ha'omer.* It is well known that the job of *Sefirah* is to count the days leading up to Shavuos in order to be properly prepared to receive the Torah anew. But how does counting *Sefirah* prepare us for that great event? Many of the commentators take the path of explaining that *Sefiras Ha'omer* is supposed to drill a certain message into us so that, after forty-nine nightly exercises, we will be ready to receive the Torah. What is that message that we must imbibe? Perhaps the answer can be discovered if we understand what the *Omer* has to do with *Sefirah.* Yes, this latter sentence was very enigmatic. Let's take the next step in our journey by explaining what it means.

The halachah states that in Eretz Yisrael (and according to many authorities, even in the Diaspora), the new crop of grain (*chadash*) cannot be used until the sixteenth day of Nissan passes. After that, the new grain is deemed "old" (*yashan*) and may be eaten. In the time of the Beis Hamikdash, the transition was made with a special offering called the *Korban Omer.* So essentially, when we start counting *Sefirah* from the sixteenth of Nissan, we are in reality counting from the day of the *Korban Omer*—hence the term *Sefiras Ha'omer.*

This piece of information is valuable to our quest. We now know that to properly prepare for Shavuos we need to work on imbibing the message of the *Korban Omer.* But what is that message?

Let's take a look at the Maharal's exposition of the mitzvah of *Sefirah.* In his *sefer Drush al HaTorah*, the Maharal writes as follows: *The mitzvah of* Sefirah *is to prepare a person to accept the yoke of Torah and mitzvos… Man was created to toil…*Sefiras Ha'omer *leads us to* Mattan Torah *because it prepares us to accept the yoke and it prepares us to toil.*

Even without reading any more of the Maharal's words, we already have gleaned a great piece of the puzzle. The Maharal is teaching us that the growth of *Sefirah* is to learn to submissively work hard as servants of Hashem. However, we do not yet know how *Sefirah* teaches us this lesson. Let's read on:

In order to enable us to learn the lesson of the yoke and of toil, Hashem commanded us to bring the Korban Omer. *The* Omer *teaches us this lesson in that it is made out of barley… Barley is the food of a beast, and our* avodah *in the days of* Sefirah *is to be like a beast. How so? Beasts in general (and oxen and donkeys in particular), have one purpose—hard work… We have to learn to be like these beasts and work hard under the yoke of Torah and mitzvos. Hashem commanded us to bring a* Korban Omer, *made from animal food, so we will learn to work hard like a beast.*

With this Maharal, our questions about *Sefiras Ha'omer* have been answered. The job of counting *Sefirah* is to recall and imbibe the message of the *Korban Omer*—that we are Hashem's servants and we must toil to serve Him. We work on this message for forty-nine days so that by the time Shavuos comes, we will have fully accepted the yoke of Torah and mitzvos.

Now perhaps we can backtrack and answer our initial questions. Why is Yissachar aligned with Iyar? The answer is because Yissachar is the one who *bent his shoulder to bear the yoke of Torah.* It seems that it wasn't Yissachar's scholarship that made him stand out with regard to this month. It was his acceptance of the job of serving the Almighty. It was his willingness to toil for the sake of his Master. This also explains why Taurus is the icon of our month. A bull's main quality is his ability to bear a yoke and work hard. That is the beauty of a beast. He toils with all he has to serve his master. Yissachar (the strong-boned donkey) and *Shor* (the Bull) are role models we should emulate in our lives. The *avodah* of Iyar is to learn to accept the yoke of toil and to work for Hashem with all we've got. That is the message we focus on when we count the *Omer*, and that is the only way we can reach a true *Kabbalas HaTorah.*

Barley, Bagels, and Bridges

The Jewish calendar is never empty. As soon as one holiday is over, we move on to the next. Purim is immediately followed by Pesach, which in turn is followed by Shavuos. And just to ensure that we don't suffer from withdrawal in the long break between Pesach and Shavuos, the Torah provides another calendar-related observance—*Sefiras Ha'omer.*

You shall count for yourselves—from the morrow of the rest day... seven weeks...fifty days (*Vayikra* 23:15-16). It seems there is a mitzvah to count the days and weeks starting from the second day of Pesach and ending fifty days later on Shavuos. If Hashem wants us to do this, we will. But one is permitted to ponder the meaning behind the commandments. What precisely is it that we are counting? What is it about the days between Pesach and Shavuos that needs to be numerated?

In elementary school, we were always told that we are eagerly counting to the day of *Kabbalas HaTorah*, receiving of the Torah. While we should never dismiss what our teachers taught us, it would probably be criminally superficial to carry nothing but our childish understanding through adulthood. There must be more to the mitzvah. An indication of this can be found in the fact that when an event is being anticipated (e.g. a birthday or wedding) the days are counted *down*, not up. If so, shouldn't the days to Shavuos be checked off as we excitedly watch the remaining days dwindle?

A good first step would be to find a frame of reference. Where else in a Jew's life do we count days? Anyone who *davens* Shacharis (and doesn't run out before the end!) knows the answer—we count the days of the week. What is it that we are counting when we say, "Today is *yom rishon l'shabbos*"? It is a difficult phrase to translate, as the word "*shabbos*" can mean both "Saturday" and "the week." Therefore, *today is* yom rishon l'shabbos, can mean "Today is the first day of the week" or *l'shabbos* could be a reference to Shabbos Kodesh. And even if we presume the latter, we will still need to ask if *yom rishon l'Shabbos* means that today is the first of the seven days until next Shabbos Kodesh, or if it means today is one day from the previous Shabbos Kodesh.

The answer to these questions can be found in *sefer Chorev* (ch. 29, par. 214), where Rav Samson Raphael Hirsch explains what the goal is when we count the days in Shacharis: *Beginning with the first day after the Sabbath, you enumerate the days following the Sabbath thus: first, second, third day after the Sabbath, and so on. In this way, the sanctity that you acquired on the Sabbath finds its expression in your mode of life in the weekday, and indeed, the Sabbath itself finds its consummation.*

It seems that Rav Hirsch is understanding the counting of the days during our prayers *not* as an anticipation of a future event but rather as an effort to maintain a past event. The counting is an effort to recall and internalize that which was learned and gained on the previous Shabbos and to bring the holiness into the more mundane days of the week. In this way, Rav Hirsch says: *Sabbath itself finds its consummation.*

However, one can still ask what message we should be taking from Shabbos into the week. Let us jump backwards to chapter 21 of *Chorev*, where Rav Hirsch writes about why it is forbidden to do *melachah* (creative work) on Shabbos: *Man was…summoned to…govern the Earth. But what would safeguard Man…that he would not forget G-d…and look at the world as…his own property? How… with his controlling power…would he not come to view himself as the master? Behold! G-d crowned His work with a seventh day…that through it, Man would...be reminded to serve as G-d's servant… Six days he can fashion all things to his purpose… On the seventh day he is forbidden to fashion anything… In this way he acknowledges that he has no authority over the world…for everything belongs to G-d.*

We learn from here that Shabbos is an incredibly important tool that we must use to keep our hearts and minds turned in the right direction. But it is not enough to learn these ideas once a week. Shabbos is the focal point of these lessons, but during every day of the week we must remind ourselves of them. Shabbos is the day to contemplate these matters all day, but during the week as well we must take a moment to recall these important ideas before we start each day.

Now let's apply these ideas to *Sefiras Ha'omer.* According to Rav Hirsch's explanation regarding Shabbos, we can understand that the purpose of counting the days of the *Omer* is to remind us daily of the messages we previously learned and need to ingrain. Indeed, Rav Hirsch writes (ch. 29): *On the day after the first day of the Passover festival, you begin to count the days as follows: first, second, third day of the* Omer*—in other words, as many days after the Passover holiday. Thus, you* ***treat Passover as you do Shabbos.*** *Enumerate seven such weekly periods and…on the day after the seventh weekly period you celebrate Shavuos…****thus linking Passover and Shavuos****.*

From Rav Hirsch's comparison between Pesach and Shabbos, it seems that his intention is to imply that the goal of counting the days of *Sefirah* is to recall and internalize the lessons learned on Pesach.

Incidentally, this idea fits very nicely with the *Sefer Yetzirah's* alignment of Iyar (the month of *Sefiras Ha'omer*) with the letter *vav.* We know that *vav* is the letter of connection. As a *vav hachibbur,* it serves as a prefix meaning "and," and it translates to mean "hook" when spelled out as a word (see *Shemos* 27:10). The *avodah* of Iyar that we accomplish through *Sefiras Ha'omer* is not an isolated one. It is a bridge (the *vav hachibbur*) meant to connect point A and point B.[1]

So now all we need to discover is what message we learned on Pesach that we are enjoined to internalize over the following forty-nine days.

I think we can suggest two possible answers. Our first explanation comes from an understanding that the mitzvah is *al Sefiras* ***Ha'omer***—to count the *Omer.* What is the *Omer*? If you

---

1. See *Pri Tzaddik* (*Ma'amar L'Chodesh Iyar, os* 5): *Iyar was created with the letter* vav, *for Rav Hai Gaon writes that the* vav *connects. This is seen in the verse* (*Shemos* 27:10), vavei ha'amudim. *Additionally, the shape of a* vav *is that of a hook. Thirdly, the letter* vav *is a conjunctive prefix, used for joining words. So too, Iyar joins Nissan and Sivan so that we can merit to acquire* b'kinyan *what we saw in Nissan* b're'iyah.

look in the Torah, you'll find that the *Omer* is a *korban* (offering) of barley brought from the first sheaf of the new crop. It is brought to the Beis Hamikdash, where it is waved before Hashem. This waving permits the new crop to be eaten, changing its status from *chadash* (new) to *yashan* (old). We start counting the days from the day this *korban* is brought. *You shall count for yourselves—from the morrow of the rest day, from the day when you bring the* Omer *of the waving—seven weeks, they shall be complete. Until the morrow of the seventh week you shall count, fifty days…* (*Vayikra* 23:15-16).

What is the reason for this mitzvah? Let's take a look at chapter 42 of *Chorev*, where Rav Hirsch explains the mitzvah of *chadash* and *Omer*: *[A risk of success is that once G-d grants you financial achievement,] you will turn your glance from on High, and instead of fearing G-d, you will idolize yourself. You will come to only seek your own interests and put your gifts to selfish use… So that we don't forget that this world belongs to G-d…we have the laws of* chadash…*which are concerned with man's possessions…. One may not eat from any of the five grains of the new crop until you have waved [the first sheaf as the* Korban Omer*]…before G-d, thereby expressing that you acknowledge that all the blessing has come to you from Him.*

Is this not a tremendous lesson? How often have we fallen into the trap of thinking that we are the masters of our lives and that what we gain through the sweat of our brow belongs to us? How often do we forget that, as it says in *Tehillim* (24:1), *the Earth and its fullness belong to G-d*, and we are but beneficiaries of His kindness? I think that with all these pieces from Rav Hirsch we should be able to put forth the following idea: Every day we count the *Omer* in an effort to remind ourselves that our daily bread is from Hashem alone. Perhaps, over the span of seven weeks, we will be able to internalize this idea.

This leads one to ask how one can focus on this lesson nowadays. After all, we no longer have a Beis Hamikdash in which we can bring a *Korban Omer*. Also, even though the laws restricting the new crop are still in effect (since the date of the *Korban Omer* has replaced the *korban*), not everyone has the custom to be stringent regarding *chadash* outside of Eretz Yisrael.

So how can we study this lesson off of which *Sefiras Ha'omer* builds?

If I can possibly add something of my own, I would like to suggest that the matzah that we eat on Pesach contains a similar message to that of *chadash*. No leavening agent is used in matzah; thus, it is a humble creation. When we eat matzah, we try to eliminate the hubris that normally accompanies human involvement. Matzah is supposed to turn our hearts and minds toward the Creator in an acknowledgment of His Kingship. This is a lesson that we can learn even in exile! The message that Hashem's Kingship is contained in the matzah should be thought about each night when the day is counted. In this way we prepare for Shavuos.

As the Ramban says in *Parshas Bo*, the Torah and mitzvos are directed so that we will come to acknowledge our Creator. If we want to gear up for the day of the Receiving of the Torah, we must link it to the lessons of *chadash*, the *Omer*, and matzah.

Let's conclude with a second possible lesson we can imbibe through *Sefiras Ha'omer*. Rav Hirsch explained that *Sefirah* is the "*vav hachibbur*" (connecting *vav*) between Pesach and Shavuos. How is one supposed to acquire the Torah on Shavuos? Well, *Pirkei Avos* states: *This is the way of the Torah: bread and salt you shall eat…* Rashi weighs in here and states that the intent of our mishnah is not completely literal: *[It is not strictly necessary to live on bread and salt alone. But one should be the type of person who,] even if all he has is bread and salt, he would not be held back from involvement in Torah.*

I think that this idea is one that can be gleaned from Pesach. Throughout the year, we indulge in cakes and cookies and all kinds of fine foods. However, the strict laws of Pesach against leaven in any shape or form necessarily prevent us from indulging in all of our favorite breads, cakes, and cookies. But guess what? We survive. It's not so bad.

It's not a crime during the year to indulge a little. But it's beneficial once in a while to abstain for a few days and realize that these physical indulgences are not so essential in our lives.

We can live without them. If we can remind ourselves nightly that we really don't need to be so involved in the physical and actually work to implement the fresh perspective we learned on Pesach that we can survive with less, then perhaps by the time Shavuos comes along we'll be that type of person who, *even if all he has is bread and salt, he would not be held back from involvement in Torah.*

A Formula for Love?

The seven weeks of *Sefirah* are a time of national excitement as the Jewish people count the days leading up to the receiving of the Torah. However, joyous as this time may be, it is also a time of mourning. This is because our Sages decreed a month-long period of grieving for the 24,000 students of Rabbi Akiva who died in a plague during these weeks. This mourning entails, inter alia, a prohibition against taking haircuts, getting married, and listening to music. Thus, at the same time that we anticipate the upcoming celebration of the Torah, we also lament the tremendous loss we suffered when an entire generation of outstanding Torah scholars was lost from our people.

Consistent with the Torah viewpoint that nothing occurs in this world without first being decreed on High, the Talmud explains what the grave transgression of these students was to have warranted such a harsh punishment: *Rabbi Akiva had 12,000 pairs of students...and they all died within one time period because they did not show due respect for one another* (*Yevamos* 62b). Although nobody would ever attribute grossly inappropriate interpersonal conduct to the holy students of Rabbi Akiva, our Sages were nevertheless able to discern a small degree of civil malfeasance within these holy men that, by Divine decree, rendered them unfit to be a link in the chain of our *mesorah* (tradition).

It seems then that during this month we should be working on the 219th mitzvah—*v'ahavta l'rei'acha kamocha*—to love your friend as yourself (*Vayikra* 19:18). However, to be honest, this commandment is quite baffling. How can the Torah command a person to love somebody? One can easily claim, "If I already love him, then the mitzvah is superfluous. And if I don't love him, I can't do anything about it. I can't make myself love him!" Love is an emotion, not an action. Ostensibly, you either feel it or you don't. However, it is axiomatic that if Hashem commands us to do something, there must be a way to go about doing it. Therefore, there must be a formula that one can follow to engender within himself a love of others.

As Voltaire famously said, "If you wish to converse with me, define your terms." Ergo, to discover a formula for love, we must first define love. Ask somebody if he loves his wife. If he answers in the affirmative, ask him why. Most likely, he will begin to list for you all of her good qualities—she's so dedicated, she's so caring, she works so hard for our family, etc. What would happen then if you asked this individual, "But doesn't she have bad qualities A, B, and C?" If he is honest, he will be forced to concede that she does. Thus, we see that in spite of his wife's negative qualities, he still loves her. Now ask somebody why he hates his sworn enemy. Chances are, he will rattle off a whole slew of negative qualities from which this individual suffers—"He's stingy, he's grouchy, he has a temper, etc." Yet if you ask this man if his nemesis has good qualities A, B, and C, he will likely be compelled by honesty to admit that you are correct. Remarkably, it seems that someone can have two relationships, both consisting of an awareness of his fellow's positive and negative qualities, and yet one is a hateful relationship and the other loving. If this is so, then from where does the hate or love originate?

Rav Noach Weinberg *zt"l* once gave a brilliant answer. Love is not defined as *knowing* a person's good qualities. Rather, love is *identifying* somebody by his good qualities. The reason why a man loves his wife in spite of her bad qualities is because when he thinks of her, he thinks of her good qualities. He identifies her by them. True, she may have negative qualities, but he views them as incidental to who she really is. Conversely, when somebody thinks about his enemy, his first thought is about the negative qualities in that man. Any attendant good qualities don't immediately rise to the surface because, in this hateful man's mind, they don't make up his foe's true essence.

Now we have defined our terms and we have our formula. If one wishes to love somebody, all he needs to do is learn to *identify* him by his good qualities. Obviously, this is easier said than done. Some people have bad qualities that are very glaring while their good qualities are somewhat harder to discover. And even when we know of somebody's good qualities, it's difficult to mentally bind that person to his good traits so tightly that we automatically associate one with the other. Appreciating this difficulty, Rav Avigdor Miller once suggested the following strategy: *If you want to love Am Yisrael, you have to work on loving Am Yisrael. You must persuade yourself... Chazal say:* The mind follows the way you talk. *Say the words and let the thoughts gradually enter your mind, little by little* (*Q&A, Thursday Nights with Rabbi Miller*).

Rav Miller means that *of course* it's difficult to identify somebody by his good qualities. But you have to work on it. Find one good quality and talk about it. Mention it—mentally and verbally. Compliment your friend on it—to him, to others, and to yourself. Talk the talk, and over many years you will come to love your fellow man more and more.

In summation, we have our work cut out for us. When the Sages decreed a month of mourning for the students of Rabbi Akiva, they didn't only want to commemorate the loss that we suffered. Rather, they intended for us to learn from the mistakes that were made then and resolve to fix them as much as we can. After all, *Sefiras Ha'omer* is the time during which we prepare for *Kabbalas HaTorah*, receiving of the Torah. Hillel said that, on one foot, the entire Torah consists of the mitzvah of *v'ahavta l'rei'acha kamocha*—to love your friend as yourself—and that the rest is commentary. What better way can there be to prepare for receiving the Torah than to work on this mitzvah? It is a long process. It can take many years. But let's start today. Look for the good qualities in others. Identify them. Talk about them. Persuade yourself of their veracity. Slowly convince yourself that your friend is defined by these qualities and his bad qualities are incidental.

In this fashion, we will learn to love our fellow Jew. And just like any parent who rejoices when he sees his children lovingly getting along, may we merit to make Hashem proud of us.

Sivan

A Tale of Two Sea Merchants

On first thought, it would seem pretty easy to discover the unique *avodah* (job) that we are supposed to focus on during Sivan. Sivan is the month of *Mattan Torah*, the giving of the Torah, so ostensibly our goal should be acquiring the Torah. However, I think there are two fair objections to this suggestion. First, we just spent the entire month of Iyar preparing for *Mattan Torah*. We just completed *Sefiras Ha'omer*—which is essentially a forty-nine-day process of acquiring the Torah. So why do we need another month? Secondly, Shavuos is on the sixth day of Sivan. If the month's work is to lead up to this goal, what are we to do for the remaining twenty-four days of the month? There must be a different job on which we must focus.

Let's adopt the Jewish technique of answering a question with a question. According to the alignment of the Arizal, Sivan is associated with Zevulun. Why is this so? One would have thought that Zevulun would be the *least* likely candidate. Sivan is the month of the Torah, while Zevulun was not a scholar but rather a sea merchant. His brother Yissachar, the Torah scholar, seems like a better choice. Even if one would argue that Zevulun earned reward equal to Yissachar's (being as the former supported the latter), it is still unclear why Zevulun would be a better candidate than his learned brother to be aligned with the month of Sivan.

Let's throw in one last question. The month of Sivan is aligned with the *mazal* of *Te'umim*—Gemini, the Twins. What do twins have to do with the month of receiving the Torah?

In order to answer these questions, let's cite a midrash (*Tanchuma, Tzav*): *In the future, Hashem will come to judge Edom. Eisav (the progenitor and personification of the kingdom of Edom) will wrap himself in a* tallis *(as if to portray himself as righteous) and sit beside Yaakov.* The midrash describes how Hashem will see through the façade, judge Eisav for the wicked man that he is, and condemn him to the nethermost pit.

Isn't this an enigmatic midrash? What does Eisav hope to accomplish by dressing himself up like Yaakov? Does he really think he can fool the Almighty into thinking that he is equally righteous as Yaakov Avinu? What type of game is this?

I would like to suggest the following explanation. Eisav is not putting forth a foolish or juvenile claim. Rather, his argument is, "Master of the world! Why should Yaakov be granted eternal reward and not me? Are we not basically the same? Most people do mostly the same things most of the time. Jews, gentiles, we're pretty much the same. We both get up in the morning, wash up, eat breakfast, and head out to work. We both work most of the day, we take our lunch and coffee breaks and come home. We both get married and raise families. We spend time with our kids and work to pay the bills. We both do our daily exercises and go to sleep at night. So if we basically live the same life, why should he get eternal reward and not me?" Essentially, Eisav is wrapping himself in the same garb as Yaakov (i.e. a *tallis*) to display to G-d how his life appears to be identical to that of the righteous Yaakov.

Now, this is a much harder argument to refute. Why isn't Eisav correct? Don't the lives of a tzaddik, a righteous person, and a *rasha*, a wicked person, seem pretty similar? (Before reading on, take a moment to think about it. Can you answer the question?) The answer is that there is a very big difference between Eisav and Yaakov. Eisav lives his life this way because he wants to; he is serving himself. Yaakov, however, is serving Hashem. When he gets up in the morning, he doesn't wash up because his hands are dirty; he washes his hands because Hashem told him to. He doesn't eat breakfast because he's hungry; he eats so he can have strength to accomplish whatever Hashem sends his way that day. He doesn't go to work because he wants cash; he does so because he wants to be able to do what Hashem wants him to do—pay tuition and other bills, give charity, buy mitzvah objects, etc. He doesn't marry because he wants a pretty wife; he marries because he wants to raise a family of Torah-true Jews. And lastly, he doesn't sleep at night because he's tired; he goes to sleep so he can be refreshed to serve Hashem the next day. So while Eisav and Yaakov may appear to be identical, they are in reality as different as can be.

This is the reality of twins as well. While twins may be identical in outward appearances, they are almost invariably of different mindsets and outlooks. It would be a great mistake to assume that twins are the same person. Let's apply this idea to Zevulun. He is a sea merchant, and every day he goes to work with his friend Bob the sea merchant. They do the same thing all day, every day. So what's the difference between Zevulun and Bob? Zevulun will bask eternally in the radiance of the *Shechinah* (G-d's Divine Presence) and Bob won't. How does Zevulun accomplish this? The answer is—through the Torah. This is the great trick of the Torah. Through it, one can change his life from mundane to holy.

Perhaps this is the *avodah* of Sivan. We already worked to acquire the Torah. Now let's use this acquisition for what it was intended. Living life for the Torah doesn't happen automatically. Au contraire, a Jew's life can look identical to the life of his gentile neighbor. The major difference often lies in the intent behind the deeds. Zevulun and Bob are like twins; they lead identical lives. But Zevulun is the greatest representative of Sivan because he uses the Torah to direct his *kavanah* (intentions) toward Hashem, thereby making his mundane life one of holiness and virtue. The *avodah* of Sivan then can be described as that of having the proper intent in all that we do, in order to achieve a life of holiness.

To Be a Twin

Sivan is a wonderful month, to be sure, as it is the month during which we receive the Torah on the holiday of Shavuos. However, since Shavuos takes place at the beginning of the month, a growing Jew might just be left wondering what his *avodah*, specific job, is for the remainder of the month. What does Hashem want us to accomplish?

It is well known that the *avodah* of a month is represented by its *mazal.* The constellation of Sivan is Gemini, commonly referred to as "the Twins." Who are these twins and what are we supposed to learn from them? If we can answer this question, we will know our job for this time of the year.

In order to explain the lesson of the Twins, we'll need to take an extended detour and analyze the holiday of the month. The understanding we gain will help us truly understand what Sivan is all about.

As we mentioned before, Shavuos is the time when Hashem gives us the Torah. What exactly is this thing that we call "Torah"? Some of our unaffiliated brethren might say it's a cultural document. Discounting this ludicrous option, a more connected Jew might say it's our history. This too is largely inaccurate, as the historical aspect of the Torah is supremely peripheral when compared to its true purpose. A more knowledgeable Jew might say the Torah is a set of laws. Now, at least, we are getting closer to the truth, but even this definition is largely deficient.

The proper answer can be found when we note the meaning of the word *torah.* Quite literally, it means "instructions." The Torah is primarily an Instruction Manual. However, this piece of information is not enough on its own. What is the Torah coming to instruct us about? Again, we will delineate the options: Maybe it's instructing us on how Hashem wants us to live. This is true. But *why* does Hashem want us to live a particular way? Some will say it's because the Torah way of life will make us holy. Again, this answer is true. But what exactly is this thing called holiness and why is it so special?

Let's turn to the *Derech Hashem* (part 1, ch. 2), where the Ramchal discusses Hashem's purpose in creating the world:

Hashem's will is to bestow good on others. However, since Hashem is perfect, it would not suffice for Hashem to bestow anything other than the ultimate goodness. Yet this [ultimate goodness] is only found in Him.

Therefore, Hashem's desire can only be satisfied when He imparts the goodness of Him Himself. The reality of true goodness therefore is the state of deveikus*—clinging to Hashem Himself.*

This incredibly fundamental piece is saying that the purpose of our existence is to cling to Hashem—because only that is considered true goodness.

However, clinging to Hashem is easier said than done. What does it mean to "cling" to an infinite, non-corporeal Being? To physical beings we cling with physical arms, but what does it mean to cling to something spiritual? A complete exposition of this discussion is beyond the scope of this essay, but in one word, the answer is "similarity." Spiritual clinging can only be accomplished when we are similar to the one to whom we wish to cling. This is because the world of spirituality has not one of the physical components or demarcations that we find in the physical world. The only descriptions that exist in the spiritual world are conceptual, so the only way spiritual proximity can be found is through conceptual equivalence. Therefore, only if we are similar to Hashem are we able to partake of that ultimate goodness that He Himself is.

So now we have the definition of holiness—similarity to Hashem. We also now know why it is desirable—because it's the only true goodness. But how does one become similar to the Infinite Creator? The answer is, through the Torah. The Torah is an Instruction Manual that instructs us in this process. While it is obviously impossible to fully comprehend the essence of our G-d, through the Torah we can learn how to become as similar to Him as we possibly can.

However, even with this clarity, we still need to delve further. How does one use the Torah to become similar to Hashem? How exactly should one utilize the Manual to achieve this goal? Let's turn back to the *Derech Hashem* (part 1, ch. 4): *Man's spiritual soul is mired in the physical body. But precisely through the body, the soul can acquire [spiritual] nobility and honor... What this means is that Hashem commanded man to carry out certain procedures...and when man makes use of the [physical] world and carries out those procedures...*

the physical action will incorporate the perfection [of the ultimate goodness of Hashem Himself] into the one who does the procedure.

In this piece, the Ramchal is explaining to us what exactly Hashem has in mind when He commands us to do His mitzvos. Simply put, mitzvos are physical actions that "magically" make us more similar to Hashem! Sometimes this is easy to understand. We all know that giving charity makes us more generous and that acts of kindness make us more kind. Other mitzvos are harder to fathom. How does shaking the Four Species on Sukkos or eating only kosher food make us similar to the Creator? The answer is that we don't always know, but what we *do* know is that each and every one of the 613 physical procedures somehow makes us more similar to the Holy One.

It would seem that we've just about come to a complete understanding of the purpose of the Torah. However, this discussion would be incomplete if we didn't also consider the concept of *limud*—studying the Torah. Torah can't possibly be only about fulfilling the mitzvos, because we know that we are even supposed to study the parts that cannot be fulfilled for one reason or another. What is the objective of Torah study? If we were to follow our earlier assumption that the Torah's purpose is to make us more similar to Hashem, then we must ask how *limud haTorah* accomplishes that goal. The answer is that studying the Torah makes us think like Hashem—as much as we are possibly capable of doing so. Torah study teaches us His views and approach. When we contemplate His words, we learn His perspective on everything. While mitzvos align our hearts to be like Him, Torah study straightens out our minds to think like Him.

By now it should be eminently clear why Sivan is matched up with the Twins. Sivan is the month of the Torah—whose foremost purpose is to make us and Hashem into "twins."[1] The

---

1. This idea must be taken with all the respect with which it was written. Obviously, it is impossible to truly become a "twin" with Hashem. However, we all know (as the entire essay describes) that we are supposed to work to become as similar to Hashem as possible. It is in that light that we attempt to

Torah is given to us so we can align our hearts and minds with the Holy One. In Sivan, we take the Torah we were given and use it to become similar to Hashem. May we be blessed with this ability and thereby fulfill our destiny to be a Holy People.

Waves and Weapons

For those of you who are familiar with our modus operandi of matching each of the months with one of the twelve tribes, I am sure Sivan will have you perplexed. According to the Arizal's system, Sivan aligns with Zevulun—the sea merchant. One could ask a fair question: What led the Arizal to the conclusion that Zevulun should be aligned with Sivan instead of perhaps Yissachar—the scholar who spends all his time studying the Torah that was given in this month? One might suggest that since Zevulun pays all of Yissachar's bills, he is considered Yissachar's equal. However, while it may be true that Zevulun will receive an equal portion of Heavenly reward, we still don't know why Zevulun should be matched up with the month of the Torah any more than Yissachar, who actually does study the Torah all his days.

Before we answer this question, let's add one more: The *Sefer Yetzirah* writes that every month is aligned with a particular letter—with Sivan's letter being *zayin*. In what way can we understand this?

In order to answer these questions, let's take a detour to discuss the origins of the name Zevulun. The Torah (*Bereishis* 30:20) describes how Leah stated, "*Hapa'am yiz**bil**eini ishi*—'Now [that this child is born] my husband will make his permanent dwelling with me, for I have borne him six sons.'"

---

become a "twin" of Hashem. In a different context, we even find *Chazal* using such terminology. *Pischi li achosi…samasi…—Amar Rav Yannai, te'umasi…*—When Hashem calls us *tamasi* (My perfect one), Rav Yannai says that we can also change the vowels and read it that Hashem calls us *te'umasi*—My twin (*Shir Hashirim Rabbah* 5:3).

The Torah is explaining that the name Zevulun comes from the Hebrew word *zevul*—a permanent dwelling place. This is strange, considering the fact that Zevulun is known as the overseas traveler. While a sea merchant may indeed have a house, he rarely spends time there. If anything, the sea would be called his dwelling place more than the dry land. While some may suggest that Leah's statement was of limited significance, the midrash (*Bereishis Rabbah* 71:3) debunks this theory by explaining that the names of the tribes are laden with significance and indeed define who they and their descendants are. So why is the frequent flier associated with permanent dwelling places?

Perhaps we can begin to answer this question by looking at the Vilna Gaon's explanation of *Sefer Yonah*. Although he does not, Heaven forbid, negate the historical veracity of this book, the Vilna Gaon explains that *Sefer Yonah* can also be understood as a parable to the soul's journey through the world. The traveler is the soul, the boat is his body with which he moves along, and the sea is planet Earth. *Sefer Yonah* describes how the soul often tries to flee from the Will of G-d but is unable to escape his spiritual calling. What can we draw out from this exposition? It seems that the Vilna Gaon understands that this world can be compared to the sea. So Zevulun is not one who stays *away* from the dwelling place of civilized men, but rather his abode *typifies* such places.

However, we still must wonder why this world would be considered a permanent dwelling. Does *Pirkei Avos* (4:16) not teach us that this world is but a corridor to the World to Come? What benefit could there be in assigning a tribe the job of making this world a permanent dwelling place?

The answer is that we are not commanded to make the world a *zevul* for us, but rather for Hashem. Hashem made the world so that we can earn eternal reward by making it into a dwelling place for Him. Our Sages teach (*Tanchuma, Naso* 16): *Nisaveh Hakadosh Baruch Hu lihiyos lo dirah b'tachtonim*—"Hashem desired (as it were) to have a dwelling place in the physical realm." And how can the Holy One dwell in the lower world? The answer is through the Torah. The Torah describes precisely how we must live our lives. When we follow the laws and ways of the Torah, the world becomes a Hashem-centric place. If we cause the whole world to be run in accordance with G-d's Will, then the world indeed becomes a "dwelling place" for the Holy One.[2]

However, one could still ask, why is this specifically Zevulun's job? Doesn't everyone have an obligation to make the world a G-dly place? The answer is that, of course, everyone has this obligation. But Zevulun is uniquely suited to fulfill this task because he is the one who is "out there" in the world. The businessmen who interact with the peoples of the Earth are in just the right position to show everyone what it means to live according to the Torah. Yissachar the scholar indeed makes the globe a G-dlier place by increasing knowledge of Hashem's "way of thinking." Also, Yissachar is the one who provides guidance on living properly according to halachah. His contribution cannot be understated. Nevertheless, while Yissachar's halachic renderings are truly the only way we know how to live *in* the world, it is Zevulun who actually brings those rulings *to* the world. The opportunities Zevulun has to make a dwelling place for Hashem are that much more encompassing. So, in reality, it is Zevulun who is most suited for the month of the Torah, for it is he who extends its reach the furthest.

Let's wrap up by explaining why Sivan is aligned with the letter *zayin*. You see, *zayin* is not just the seventh letter; it is also a Hebrew word which means "weapon."[3] In the month of the *zayin*, Hashem gave us a weapon with which we can fight the world war. You may not have realized we are engaged in such a

---

2. These aforementioned thoughts are extrapolated from *Pri Tzaddik, Ma'amar L'Chodesh Sivan, os* 1: *Sivan is aligned with…the left leg… But what does the left leg have to do with receiving the Torah? The answer is…the left leg symbolizes external (i.e. non-spiritual) matters…which through [Torah] even the left leg becomes [like] a right leg.* This essentially means that the goal of Sivan is to take the Torah and use it to turn our physical lives into spiritual lives.

3. This idea and the rest of this paragraph is based on *Pri Tzaddik, Ma'amar L'Chodesh Sivan, os* 1: *The power of the Torah given to us…is the power of the sword…this is why the* zayin *is aligned with Sivan, the month of receiving the Torah.*

war, but the fact is that we are. The nations of the world struggle mightily to crown Man as the king of the universe, while we Jews contend that Hashem is the King of the universe. The nations fight to create a Man-centric world, while our people battle to construct a dwelling place for the Almighty. The Torah is our only weapon, because without it we have no way of knowing how this holy abode can be built. Molding the world in accordance with Hashem's plan cannot be done simply by following our hearts. We must work to incorporate halachah into every aspect of our lives. We must act in a way befitting the bearers of G-d's Word and we must cause G-d's Name and ways to become beloved in the eyes of all humanity. All this can only be done with our secret *zayin*—the Torah. May we merit to receive the Torah with full hearts and learn to use it in fulfillment of Hashem's Master Plan.

[Author's postscript: I wrote this exposition of Zevulun based on a Rashi that explains that *zevul* is a permanent dwelling. However, I subsequently saw another one of the seventy faces of the Torah—in the Ramban. Ramban learns that the word *zevul* actually connotes a *temporary* dwelling; a shelter not meant for permanent use. In this understanding, Leah's prayer upon the birth of Zevulun was not so broad as to include a hope that Yaakov would make his permanent dwelling by her, rather that he would simply take shelter by her more often.

According to this Ramban, one can understand that Zevulun's primary place of residence—the sea—is indeed only a temporary place (as is the simple understanding). As for Hashem dwelling in our world, this same Ramban explains that proper respect for the Supreme One would dictate that we only can pray that He deign to make His temporary dwelling with us (through our secret weapon—the merit of our Torah and mitzvos). Ramban thus explains Shlomo's statement upon construction of the Beis Hamikdash (*Melachim I* 8:13): *I have built a* zevul *for You.*]

Tammuz

The Weeping Statue

Tammuz is the beginning of the sadder part of the calendar, as we enter the season of the destruction of the Beis Hamikdash. But even as we mourn for what we have lost and yearn to regain it, we cannot stop growing in our *avodas Hashem*, service of G-d. Tammuz is not an exception to the rule that every month has its own unique way in which we must come closer to our Creator. So what is the *avodah* of Tammuz?

Let's explore this topic by starting with a question: Where in Tanach does it say the word *Tammuz*? You are probably aware of the fact that the five books of the Torah do not name the months but rather number them. The names we are familiar with (e.g. Nissan, Iyar, Sivan, Tammuz, etc.) originated in Bavel. Hence, we will only find a reference to our month's name in the Prophets or the Writings.

It turns out, the word *Tammuz* is found in the book of *Yechezkel* (ch. 8), where the *navi* writes that Hashem showed him how the Jewish people had degenerated to such a degree that they brought idol worship into the Temple itself.

And [G-d] said to me, "Enter and see the evil abominations that they commit here!" So I entered and saw every sort of [idolatrous] image...all the idols of the House of Israel... [There were] seventy men, each one holding a censer [of incense that they were offering to the idols]. The *navi* also describes the sun worshippers: *At the entrance...there were some twenty-five men...with their backs to the Sanctuary...bowing eastward to the sun.*

Thus far, we see all sorts of idol worship with which we are at least somewhat familiar. But now let's look at verse 14: *[G-d] brought me to the gate...in the north, and behold, there were women... causing Tammuz to cry.* What is this supposed to mean? Isn't

Tammuz a month? How can it cry? And what does this have to do with idol worship?[1]

Rashi steps in and explains that Tammuz was the name of an idol. The female priests would heat up the Tammuz until it was so hot that its lead eyes would melt, thus creating the impression that the idol was crying and begging to be given sacrificial offerings. While this is definitely interesting, there are two major questions that need to be addressed: The first question, we alluded to earlier. Namely, the fact that our month and this deity share a name indicates that they share a commonality. What is that commonality? The second question is that the verse describing what Yechezkel saw regarding the Tammuz seems to be incongruous to what he was describing in the rest of the *perek*. As quoted above, Hashem showed Yechezkel all the idols that were *being worshipped* in the Temple. Hashem showed the incense being offered to the statues and how people were actually bowing down to the sun. However, we don't see any description of the Tammuz *being* worshipped. We see how the women would work to create the illusion of the idol crying *to be given* service but the verse does not describe the ensuing service occurring. Why is this pre-idolatrous act—wrong though it definitely was—included in Hashem's list of abominations that caused His Presence to depart from the Temple in a way that the worship and pre-worship are seemingly equated?

In order to answer these questions, let's take a detour and explore a different aspect of the month. As we know, each month is aligned with one of the tribes of Israel, with Tammuz being aligned with Reuven. What are Reuven's essential and unique qualities? If we can answer this question then perhaps everything will be resolved.

---

1. I found this idea in the *Pri Tzaddik* (*Ma'amar L'Chodesh Tammuz, os* 1), where it is quoted from the Raavad's commentary on the *Sefer Yetzirah*. However, the explanation for this phenomenon given by the aforementioned commentators is esoteric and beyond my understanding. The explanation I present herein is my own, albeit based on what I hope is *da'as Torah* (the correct Torah perspective).

Let's take a look at the *parshah* in which Reuven was named, because we know that a name describes one's essence: *Leah conceived and bore a son, and she [prophetically] called his name Reuven* (*Bereishis* 29:32). Rashi explains that Reuven is a contraction of the words "*re'u*"—look and see, and "*ven*"—the difference between. Rashi elucidates by quoting the Gemara in *Brachos* 7b: *Look at the difference between my son and my father-in-law's son, Eisav. Eisav willingly sold his rights of the firstborn to Yaakov and yet he still hated Yaakov for it, while [I know prophetically that] Reuven will lose his rights of the firstborn, which will be given against his will to Yosef, and yet Reuven will not harbor any ill will toward Yosef as a result.*

This fascinating Rashi is describing the essence of Reuven. What is this *middah* (trait) that Rashi is mentioning, and why is it so special? The answer is that people have a natural but negative tendency to always want to be ahead of their peers. We like to believe that we are better than anyone else. The *yetzer hara*, evil inclination, drives us to want to push everyone else beneath us because it is too painful to imagine a reality in which somebody is doing better than us. In reality, this is a disgusting trait that we need to do our utmost to uproot. Is G-d not all-powerful? Is He not loving and benevolent? Can't He provide every single one of our needs and wants while simultaneously doing the same for our neighbor? Why do we always have to be on top and make our friend the deficient one? This tendency is illogical and evil, and yet it thumps within us always.

Reuven, being one of the twelve sons of Yaakov, was one of the greatest people who ever lived. Astonishingly, he managed to grow beyond this evil tendency and conquer his baser self. While Eisav resented Yaakov his entire life for receiving his firstborn rights, which he sold to him of his own free will, Reuven accepted the place in life that Hashem decided was best for Him. He was thereby able to maintain a loving and caring relationship with his younger brother, who became his superior.

Now we can suggest an explanation for why it was so abominable to cause the Tammuz to cry. Possibly even worse than worshipping the deity was the fact that those people were restructuring their relationship with the "Divine." They aimed to make their god the needy and deficient one and put themselves in the position of the benevolent ones. They made the Tammuz cry so they could step in and be the kindly superordinates with the ability to fill all the subordinate's needs. This is an abominable trait. Why can't we be more like Reuven and accept our station in life? Doesn't G-d love us and know what we truly need much more than we will ever know? This is the *avodah* of Tammuz.

To (Let It) Be or Not To (Let It) Be

While the summer may be a time for vacation, we all know that *avodas Hashem*, service of G-d, is a year-round duty. So in what way should we be growing in the month of Tammuz? What specific task does Hashem have in mind for us during these few weeks?

Our Sages write that the months of the year match up to the twelve tribes of Israel, with the Arizal matching up Tammuz with Reuven. This alignment informs us that Tammuz is the time to learn from the ways of this great man in one way or another.

But what do we know about Reuven? The Torah does not describe this firstborn tribal progenitor at length, and he is really only singled out in three instances. First, when Reuven saw that his mother was not conceiving anymore, he brought her a fertility-inducing plant called *duda'im* to help turn the situation around (*Vayeitzei* 30:14). In the second instance, when Rochel died and Yaakov moved his bed to the tent of Rochel's maidservant, Reuven went and moved the bed to the tent of his mother Leah (*Vayishlach* 35:22). The last time was when the brothers decided to kill Yosef, Reuven protested and convinced them to instead throw him in a pit. The verse testifies: *[Reuven convinced them to throw him in a pit because he was] intending to rescue him from their hand, to return him to his father* (*Vayeishev* 37:22).

What do these three incidents have in common? Do they inform us of Reuven's special *middah* (character trait)? If we can find a way to deduce the common idea that these instances share, we'll be well on our way to understanding the growth of Tammuz.

Perhaps we can find the missing piece of information by analyzing the one calendric observance of Tammuz—Shivah Asar B'Tammuz. The mishnah in *Ta'anis* (26b) describes how there were five tragedies that occurred on Shivah Asar B'Tammuz, which therefore make it necessary to fast and repent on this sad day: 1. In the time of the first Beis Hamikdash, our enemies laid siege to the Temple and cut off the supply of sheep for the daily *Tamid* offering. It was on Shivah Asar B'Tammuz, three weeks before the destruction, that the Temple supply ran out and the *Tamid* service came to an end. 2. In the time of the second Beis Hamikdash, an idolatrous image was set up in the Sanctuary of the Temple. 3. Around sixteen years before the Great Revolt against Rome, a wicked man named Apostomus captured and burned a *sefer Torah*. 4. It is the day Moshe Rabbeinu broke the Tablets. 5. The destruction of the Second Temple began when the walls of the Holy City were breached by our enemies on this day.

Are these five occurrences disparate events that coincidentally happened on the same day, or do they perhaps share a common denominator?

I believe that one could suggest that these events do indeed share a commonality. They were all challenges to the status quo. The *Tamid* was an offering that was brought every single day without fail. Yet on that tragic day, the status quo was challenged. The image in the Sanctuary also broke the barrier, for the tragedy was not so much that the Jewish people had sunk into idolatry (something unfortunately not new), but that they had stooped to a new low by bringing a foreign god into Hashem's house. The burning of the Torah and the breaking of the *Luchos* also clearly defied the standard, for what could possibly constitute a decimation of the standard more than the destruction of that which bears it? Lastly, can there be any greater representation of the concept of challenging the status quo than the example of city walls being breached?

Let's run with this idea. The mishnah seems to indicate that Tammuz is a time when change is in the air. Things have been a certain way, but the norm does not have to stay as it is. The status quo can be challenged. But let's ask a fair question: All the examples in the mishnah are tragedies—aberrations from a positive standard. Does that mean that this month's quality is purely negative? I assume the answer is no. All qualities can be used both for negative and positive. Obviously, if the status quo is that Hashem's Will is being fulfilled by having the *Tamid* offering brought every day, then it's a bad thing to break the standard and desist from offering it. If the city walls have been protecting the populace for many years, then it's clearly a bad thing for the existing state to change.

But what if the standard is something inappropriate? Is it proper for one to shrug his shoulders and say, "Well, that's the way things are"? Of course not! It is the mark of tzaddikim (righteous ones) to be able to step forward and say that they are willing to make a change even if it's not how things have been done thus far. Perhaps this line of reasoning would lead one to say that while one has to be on guard in Tammuz to not let any positive standards break, one also has the job during Tammuz to step forward and challenge any standard that upholds the negative.

Now let's return to Reuven. A closer analysis of the three instances in which he is mentioned reveals that Reuven excelled in this *middah* of challenging the status quo. When he saw that his mother was no longer conceiving, he went and proactively did something to change the situation. When he perceived that his mother's honor was being slighted, he challenged the status quo and moved her bed from where it was to the place where he believed it belonged. And when his brothers unanimously decided to kill Yosef, he spoke up and altered the reality of the circumstances and set a higher standard. In all these cases, Reuven challenged the status quo. (Then again, it shouldn't come as a surprise that this *middah* was strong in Reuven. After all, isn't the firstborn the one who changes a couple into parents and makes a twosome into a family?)

The conclusion we can draw from here is that Tammuz is a month of change, for better or for worse. A person has the inherent ability to challenge the way things are and change any given situation. He can break a positive standard and set new lows, or he can stand up for what's right and challenge a wrong. One can bring tragedy by lowering the benchmark for success, or he can improve the world by raising the bar. Let's learn to challenge the status quo in the proper way and be like Reuven. If we can do that, we'll have an accomplishing Tammuz.

The Three Weeks: A Time to Love

The Three Weeks make up one of the longest observances in the Jewish calendar, running from Shivah Asar B'Tammuz all the way until Tishah B'Av. Proper observance of these solemn weeks entails restricting oneself from a number of things, including but not limited to getting married, listening to music, making a celebration or party, taking a haircut, and buying new clothing.

Every growing Jew should be asking himself during this period: What were the prophets intending when they instituted restrictions during these weeks? What is it about this calendric period that necessitates putting a damper on our normally joyous lives? For those of us who attended Jewish elementary schools, the answer probably seems pretty obvious. We were always taught that this is a period of mourning over the destruction of the Beis Hamikdash.

However, one could ask a fair question on this assumption. Isn't Tishah B'Av the day of mourning for the Temple? Why did our prophets ordain that we must begin the mournful observances three weeks earlier? Again, if you attended a Jewish elementary school, you'll probably answer that it is appropriate to start mourning on Shivah Asar B'Tammuz because it is the anniversary of the day our enemies breached the walls of Jerusalem. Now, technically, you'd be correct in saying so. After all, the Gemara refers to these three weeks as *Bein Hametzarim*—"Between the Distresses"—"Distresses" referring to Shivah Asar B'Tammuz and Tishah B'Av. However, one can still wonder why such a long mourning period is necessary. It seems quite abnormal for our Sages to consecrate such a large period of time for *any* observance. Granted, we believe that all our sorrows stem from the destruction of our Temple. But even if you argue that all events that relate to said destruction (e.g. the walls being breached) should be mourned as well, one still can question the necessity of three whole weeks being dedicated to that end.

Perhaps one could suggest that if all that was necessary was the mourning of the historical tragedies that took place during this time, then truly the two days of Shivah Asar B'Tammuz and Tishah B'Av would suffice. However, our wise prophets intended much more than mourning over past glory. They knew the dictum of our Sages (*Yerushalmi, Yuma* 1:1): *Any generation that does not see the rebuilding of the Temple in their days is as if they saw it destroyed in their days.* They also knew the exposition of the aforementioned quote that states that any generation that doesn't see the rebuilding is obviously suffering from the same sins that caused the destruction in the first place (because if not, they would merit the Redemption).

Based on this understanding, the prophets saw a need to enact something to fix the problem preventing the Redemption. This is what brought about the observance of the Three Weeks. The *avodah*, specific job, of these three weeks can be understood as almost being separate from the *avodah* of mourning we practice on Tishah B'Av and Shivah Asar B'Tammuz.

This thought process leads one to realize that he has two choices during these days. He can follow the letter of the law and submit to the restrictions until the weeks pass. Or he can follow the spirit of the law as well and choose to utilize the given guidelines to fix the problems that make us need to mourn in the first place. Let's discover how our Sages intended for us to become bigger and better people.

The Talmud teaches that the destruction of the Temple that brought about our current exile was caused by baseless hatred.

Our ancestors did not properly fulfill the mitzvah of loving their fellow as themselves. Their strife and infighting caused the *Shechinah*, Hashem's Divine Presence, to depart from the Beis Hamikdash. Once the *Shechinah* was gone, the hollow Temple shell didn't stand a chance against our enemies. And, as we mentioned earlier, if the Beis Hamikdash still hasn't been rebuilt, that means that we too have yet to purge ourselves of baseless hatred.

What still needs to be addressed, however, is *how* exactly our Sages wanted us to work on this issue. In what way can the restrictions they enacted help us learn to love our neighbor as ourselves? To answer this question, let's take a look at the *Kli Yakar* (*Vayikra* 16:30) that explains why Moshe gathered Bnei Yisrael to build the Mishkan specifically the day after Yom Kippur:

Only when they were a unified congregation…would Hashem accept… [them]. [But] it is normal for people to beg forgiveness from their friend on Yom Kippur and then...go back to their evil ways. So Moshe gathered them [immediately] after Yom Kippur and unified them into one congregation. This [immediacy was a necessary ingredient] because all strife stems from interpersonal dealings…

This enigmatic *Kli Yakar* is explaining that baseless hatred stems from everyday transactions and interpersonal dealings. It seems that one can get so caught up in all his work that he grows self-centered and, consequently, argumentative. Moshe Rabbeinu knew that the best chance he had at uniting Bnei Yisrael into a unified whole was immediately following Yom Kippur, because Yom Kippur is the day when everyone removes themselves from indulgences and self-service.

Perhaps now we can suggest an explanation for what the prophets intended with the restrictions of the Three Weeks. Perhaps when we were told not to make weddings, listen to music, throw parties, take haircuts, or buy new clothing, we were really being told to stop focusing so much on ourselves. It's possible that the Sages were telling us to actively practice not being so selfish and instead think of others more often. Maybe they wanted us to come across a certain pleasure during this time period and, instead of indulging in it, to go give joy to somebody else. If we can lessen our self-love and increase our love of others, then maybe this year we'll merit the rebuilding of the Beis Hamikdash.

Av

Can You Loan Me a Letter?

As Tammuz ends and Av begins, we mark the calendric observance commonly known as the Three Weeks. This is the sadder part of the calendar, as we mourn the destruction of the Beis Hamikdash and all the tragedies that occurred as a result. We all know that Hashem decreed our present exile because of *sinas chinam*—the baseless animosity and dislike we bear toward our Jewish brothers and sisters. How can we rectify this problem? It is deep-seated, but with a little analysis we will hopefully discover a tool to lessen this brotherly distaste and thereby bring the Redemption closer.

The *Sefer Yetzirah* writes that each month has its own letter, with Tammuz and Av matching up with *ches* and *tes*, respectively. Although each letter represents a force that can be bad or good, the combination of these letters does not bode well—as alluded to in the fact that together they are the root letters of the word *chet*—sin. *Bnei Yissaschar* (*Ma'amarei Tammuz V'Av* 1:2) explains that the fact that these months' letters combine to spell "sin" should not come as a surprise, since we know that the tragedies that occurred in these months were a result of our sins. So let's search for a message of the *ches* and *tes* combo and see if we can use it to turn these months around.

The source we will turn to is the *Da'as Zekeinim* in *Vayechi*. The verses there describe how Yaakov called his sons together shortly before he died to bless them and to reveal the date of the final Redemption (the *Keitz*). When Yaakov tried to reveal the date, the Divine Spirit departed from him and he was no longer privy to the secret information. The verse then describes how Yaakov proceeded to give the blessings. *Da'as Zekeinim* elucidates this narrative by recording an exchange that took place between Yaakov and Hashem after Yaakov's failed attempt to reveal the *Keitz*.

When Yaakov saw that he was denied permission to reveal the Keitz, *he feared that his children were unworthy. Perhaps they had sinned. Yaakov then reasoned that this could not be for "they do not have the letters of* chet *on their name." Meaning, not a single one of the tribes had a* ches *or a* tes *in their names. Yaakov reasoned that since one's name reveals his essence, the essence of the tribes must be pure because they had no letters in their names that could represent sin! However, Hashem denied Yaakov permission to reveal the* Keitz *by countering, "But they also don't have the letters* kuf *or* tzaddi *in their names (these letters spell the word* keitz*). They do not have permission to know this secret."*

Before analyzing this *Da'as Zekeinim*, let's make a disclaimer: The essences of letters and names are Kabbalistic in nature and consequently out of reach of the vast majority of us. Therefore, any questions we pose may be off base, out of context, and/or completely wrong. With this shot of humility, let us delve in as far as we are able.

If we run with the premise that one can only receive what his essence (reflected by his name) allows, then how was Yaakov able to bless any of the tribes? The word for blessing, *brachah*, is spelled *beis, reish, chaf, hey*. However, if you look at the names of the twelve sons, not a single one of them has all four! While Reuven had a *beis* and a *reish*, wasn't he lacking *chaf* and *hey*? And while Yehudah might have had a *hey*, he didn't have the other three letters. So how could they each receive a blessing if their names didn't allow it?

To answer this question, let's take a look at the Vilna Gaon's writings on *Mishlei*. In the final chapter of *Mishlei*, we find the song of *Eishes Chayil*. Simply understood, this ballad is an ode to the Jewish woman—who is valorous in her great role of wife and mother. However, each part of the Torah has infinite levels, and through the eyes of our commentators we can glimpse the profundities of Hashem's Word. Case in point, the Gaon explains that *Eishes Chayil* is also a song of praise for the Torah. The Gaon expounds on how "*chayil*" has a numerical value of forty-eight—which is the precise number of steps one must take to acquire Torah. With this understanding of the subject of the song, the Gaon explains the verses quite differently than they would be understood if they were only referring to the Jewish wife and mother.

Let's skip to one of the last verses: *Rabbos banos asu chayil, v'at alis al kulanah.* If referring to a woman, this verse means: "Many daughters have achieved great valor, but *you* rise above them all." The Gaon, however, elucidates as follows: Rabbos banos—*only through many scholars;* asu Chayil—*can Torah, which is acquired through forty-eight steps, be acquired;* v'at alis—*and you, Torah, can only be acquired;* al kulanah—*through all of these scholars combined together.* The Gaon then explains: *There are forty-eight steps needed to acquire Torah. But it is nigh impossible for any one individual to achieve them all on his own. So each person achieves a few, and the Torah can be acquired through the sum of their combined efforts.*

Let's spell out the mind-blowing idea we see from the Gaon. It's not necessary for each individual to have every single good quality to be able to share in the *brachos* that good traits bring down from Heaven. We Jews share a soul connection, and if *you* have a certain quality, then *I* share in it as well! You contribute something, I contribute something…and between the two of us, we have everything!

I think that now we can answer the question we asked about the tribes. It was not necessary for Reuven to have all the letters of the word *brachah.* It's true that he didn't have a *chaf* and a *hey,* but he didn't need it—because Yehudah had a *hey* and Yissachar had a *chaf.* Binyamin had a *beis,* and he was able to borrow the *reish* from Reuven, the *chaf* from Yissachar, and the *hey* from Yehudah. Collectively, they had it all!

With this understanding, perhaps we can backtrack and discover a way to lessen *sinas chinam.* It's true that no group or individual in Klal Yisrael is the same as me. But that group or individual contributes something to our collective national whole that would be missing otherwise. One group is very careful about *davening* at the proper time. Other groups are very meticulous about preparing properly for *davening.* Together, we have a perfect *davening.* Some individuals are very serious and focused about life. Others are carefree and happy-go-lucky. We can all contribute something to the greater whole.

So when you feel that niggling feeling of dislike for that other person or group, try to stop and think what good they are contributing to our nation to make it more complete. In this merit, may we *all* greet the Mashiach soon.

The Ventriloquist Lion

Av is definitely a month in which we want to improve ourselves. What area of *avodah* (specific job) does Hashem want from us during these weeks? Let's see if we can figure it out.

The month of Av is aligned with the tribe of Shimon; let's analyze his essence as alluded to in his name. Upon giving birth, Leah states, *"Hashem has heard that I am hated, and He gave me even this one"* (*Bereishis* 29:33). The *pasuk* is explaining that the root of the name Shimon is from the word *shmi'ah*, which means listening. Apparently, the *avodah* of Av is related to listening.[1] How can this be understood?

Additionally, every month has a constellation, and Av is aligned with *Aryeh*—Leo, the Lion. The metaphor of a lion is found in many places in the Torah, and the significance of Av's alignment with the king of the beasts can consequently be very great. However, it may be prudent to limit our study to two specific instances.

One obvious example of one who may be compared to a lion is Hashem (see *Hoshe'a* 11:10). This anthropomorphism is logical, for there is no greater leader that can be compared to the great and mighty king of the beasts like the King of kings. However, the *avodah* of Av presumably is not to emulate Hashem's lion-like qualities—because it is impossible to obtain even a fraction of Hashem's power. So let's move on to an alternative option. We find in *Sefer Yirmiyah* (4:34) that Nevuchadnetzar, the mighty ruler of the globe-spanning Babylonian Empire, was also

1. This fact is also borne out from the *Sefer Yetzirah* where it is written that the faculty of Av is *shmi'ah*.

compared to a lion. Now, this choice strikes a chord, as Av is the month in which Nevuchadnetzar destroyed the Beis Hamikdash. Ostensibly then, there is a connection between Av and Nevuchadnetzar. But perhaps this option also should be discounted—because what could we possibly want to learn from Nevuchadnetzar's evil ways? So that still leaves us with the question: Which lion is Leo?[2]

Let's take a detour. The Rema (*O.C.* 476:2) writes that some have the custom to eat an egg (which is a symbol of mourning) at the Pesach Seder. He explains that this is because the first night of Pesach always falls out on the same day of the week as Tishah B'Av. Is this not a strange idea? It is obvious that the Rema understands that there is a connection between Pesach and Tishah B'Av. What can that connection possibly be?

Perhaps the key can be found in the Gemara in *Megillah* (14a): *Praise the servants of G-d, who are not the servants of Pharaoh.* This phrase teaches that human beings are never really their own masters. They are always either the servants of Hashem or of Pharaoh.

Hashem wants only the very best for us. As the Ultimate Giver, He wants to give us everything. If one believes that what happens in his life is solely in his own hands then he will never be happy, because he will constantly be worried about occurrences that are beyond his control. In contrast, someone who knows that Hashem is in complete control can relax in the knowledge that everything will work out in a good way.

Hashem, being the Ultimate Giver, wants us to be humble before Him so we will be able to be truly happy. But if we refuse to humble ourselves on our own, then Hashem will need to find a different way to lower our grandiose self-image. It is

---

2. The scholar might find it interesting to reference a *Pesikta D'Rav Kahana* 13:15 (also quoted in *Pri Tzaddik, Ma'amar L'Tu B'Av, os* 2) that says: *The first Beis Hamikdash was destroyed during the [month of] the* mazal *of the Lion, [destroyed] through the lion Nevuchadnetzar, but regarding the future Beis Hamikdash, the Lion—this is Hashem—will come and rebuild the Temple in the month of the* mazal *of the Lion.*

likely that Hashem will lend some of His power to a mortal so he can subjugate us and make us realize that our lives are not in our own control, but rather in Hashem's. This is the connection between Pesach and Tishah B'Av. At the Seder we eat the egg to remind ourselves that we can either humble ourselves and be servants of Hashem, or Hashem will find another way to bring us to the greatest happiness—by placing us at the mercy of a mere mortal (like He did on Tishah B'Av).

Now perhaps we can discover the identity of the Lion of Av. Of course, it is Hashem—for there is no Power other than Him. Not only is He the most powerful, He is the *sole* Power that exists. Hashem is the Lion.

However, there will be times when we refuse to humble ourselves before Him. In those cases, Hashem may decide to lend some of His power to (or more accurately, channel His power through) a mere mortal. Hashem's hope is that when humans see a fearsome leonine ruler threatening them, they might actually decide to turn their hearts back to their Father in Heaven. This is a helpful assistance that Hashem lends us when He sees we are having a hard time recognizing how little we control our lives. Nevertheless, when we encounter those powers that humble us so well, it is crucial that we realize that they don't possess any of their own power. Rather, we must get the message that we humans are all subject to the greatest Power of them all—Hashem.

By now it should be clear why Av, the month of Shimon, is the time to work on listening. When Nevuchadnetzar roared, who was really roaring? Was it not Hashem throwing His august Voice to a human dummy in order to help His children become humble people? We have to open our ears and listen carefully to discern the true source of sounds. Everything in the world happens because Hashem is causing it. World events and the decisions of world leaders are all only Hashem sending messages to those who desire a relationship with Him. If we can learn to hear the voice of Hashem when the lion's lips are moving, we'll have an accomplishing Av.

Temples, Tribes, and Timeless Deeds

How does one feel as the month of Av begins?[3] The Gemara (*Ta'anis* 26b) rules: *When Av begins, we lessen our joy*. It seems we have reached the saddest time of the year. Indeed, this month is host to a number of tragic occurrences, primary amongst them being the destruction of the two holy Temples. The above-quoted Gemara also states: *On Tishah B'Av it was decreed on our ancestors (in the Sinai Desert) that they would not enter the Land (of Israel), the two Temples were destroyed, the city of Beitar was conquered, and Jerusalem was plowed under. [Therefore,] it is forbidden [during this time of year] to cut one's hair...launder one's clothes...or eat meat, etc.*

However, Av is also host to the holiday of Tu B'Av. The same Gemara states: *[The people of] Israel had no days as festive as Tu B'Av and Yom Kippur*. It seems that Av is host to the saddest days of the year and simultaneously to one of the happiest days in the year. Tishah B'Av we are familiar with. But what is this obscure holiday of Tu B'Av?

We find that the Gemara actually asks this question: *It is understandable that Yom Kippur is a festive day because it is a day of forgiveness. But what is the significance of Tu B'Av?* The Gemara proceeds to list six joyous events that occurred on Tu B'Av. We will limit ourselves to the first two.

Tu B'Av was the day the tribes were permitted to marry into one another. [For in the generation of Moshe it was ruled that the tribes may only marry within their own tribe. But the Sages subsequently (on Tu B'Av) found basis in the Torah to say that this prohibition only applied to that generation alone.] In addition, on this day the tribes were permitted to marry into the tribe of Binyamin. [For the tribes took an oath not to marry into Binyamin following the atrocity of the concubine of Givah. But the Sages subsequently (on Tu B'Av) found a leniency based on the phraseology of the oath which rendered it non-binding on future generations.]

---

3. The entire content of this essay has been adapted from *Pri Tzaddik, Ma'amar L'Tu B'Av, os* 3.

Let's analyze these two events. While they were definitely special occurrences, why were they cause for a new holiday that would be called *the most festive day in Israel*? Were these relatively minor events really greater than the Exodus from Egypt or the Revelation at Har Sinai?

In order to answer these questions, we need to understand the idea of holidays in general. Why do we celebrate the holidays? Are they all merely commemorations of past events? The *Pri Tzaddik* explains that this is not so.

All holidays can be explained in the following way: A past occurrence happened on that day that made an impact on that calendar date. A degree of holiness was brought into that date forevermore and it is therefore appropriate to make a celebration (i.e. holiday) on that day.

Isn't this a mind-blowing idea? Rav Tzadok is saying that the event that occurred way back when caused a degree of holiness to come into the world and settle on that particular date, thereby turning it into a day of celebration! This calendar day will never be the same again because it is now a holier day—like Shabbos.

Let's apply this idea to Av. Is Tishah B'Av a day of mourning *only* because in the past the two holy Temples were destroyed? No. It is axiomatic in Jewish thought that nothing happens without a reason. Why did the Jewish people deserve to have the Second Temple destroyed? Our Sages teach that it was because of *sinas chinam*—baseless hatred. Why was it decreed on our ancestors that they could not enter the land? It was because of *lashon hara*—the slander and divisive speech of the spies. On Tishah B'Av (way back when) a contamination came into the world—from divisiveness between brothers. That impurity remains to this day—tied to our time through the calendar date. Hence, we mourn.

But all is not lost. We have the ability to bring contamination into the world, but we can also bring down purity and holiness. The corruption brought into the world on Tishah B'Av can be rectified by acts of love between brothers. If we act with *ahavas chinam*, baseless love, we can bring a G-dly light to the world and tie it down here as well.

This is what we celebrate on Tu B'Av. While the tribes at

Givah and in the Sinai Desert were certainly acting for the sake of Heaven, their actions were causing divisiveness in Klal Yisrael. On Tu B'Av, however, two wondrously joyous events occurred that enabled the tribes to marry within each other. Why is this so special? Because it brought love into Klal Yisrael. It broke down the barriers of negativity and polarization. These events created a holiness that stems from *shalom* and permanently changed the world for the better—by tying that holiness to the world through the calendar date.

Why is there no more festive day for Israel than Tu B'Av? Because it is the day when our actions caused the start of a rectification for the saddest day of the year. All our sorrows stem from a lack of Hashem's presence in the world, which in turn stems from the exile and the destruction of the Temples. We all wish for the day when Mashiach will come and rebuild the Beis Hamikdash. But it is more in our hands than we realize. Tishah B'Av shows us that we can bring impurity to the world, but Tu B'Av shows us that our deeds can bring holiness to the world and fix our mistakes. The month of Av teaches us to realize the significance of our actions.

May we merit internalizing this message, and by doing so may we see the Redemption, speedily and in our days.

Elul

Sins, Sculptors, and Circumcision

As Elul arrives, we know the Day of Judgment is coming. If our minds and hearts are awake, we are thinking about how we can prepare ourselves for the big court case. As we all surely know, the *avodah*, specific job, of Elul is *teshuvah*, repentance. We will make no pretensions about trying to compress what needs to be said on that topic into one short essay. To learn that process one needs a *sefer Sha'arei Teshuvah* and a *rebbi* (teacher). However, we *will* try to gain a Torah-true perspective that is specific to this month and will hopefully inspire us to take the next step of actually working on doing *teshuvah*.

Let's turn to our modus operandi of looking for hints in the tribes of Israel and in the *mazalos* (constellations). Since each month is aligned with a particular tribe and constellation, we can inform ourselves of the qualities of the month through them.

The *mazal* of Elul is *Besulah*—Virgo, the Virgin. Let's make the fair assumption that the maiden is the icon after whom we should model ourselves during Elul. What is special about a *besulah*? Well, we do find that our Sages use a *besulah* as the symbol of purity. She is the untouched maiden, pristine in her innocence. However, one could ask a fair question: Does Judaism really put chastity up on a pedestal? Other religions might view it as an ideal, but the Torah believes that while purity is commendable for a particular time and stage, it is appropriate and even praiseworthy to move on at a certain point. So what precisely is the message we are supposed to learn from the maiden?

Before we answer this question, let's analyze Elul's tribe—Gad. Rashi in *Vayeitzei* (30:11) quotes a midrash that describes how Gad was unique in that he was the only one of the tribes to be born circumcised. Ostensibly, this fact should immediately catapult Gad to sainthood and make him a prime candidate for "role model of the month." However, if we take a look at the

Sefer Hachinuch's understanding of the mitzvah of circumcision, we may be more hesitant. *[Why did Hashem make man incomplete (i.e. uncircumcised) in the womb? Is Hashem's handiwork deficient?] The reason Hashem created man deficient is because He wanted the perfection to be made by the hand of man himself.* We see from the *Sefer Hachinuch* that there is a great benefit to being born uncircumcised. So is there really any lesson that we can learn from the alignment of Elul with someone who *was* born circumcised?

In order to answer our questions, let's tell a little story. There was once a great sculptor who was world-renowned for his beautiful works of art. From animal statuettes to larger-than-life busts of historical figures to intricate carvings of flowers, there was nothing he couldn't make. His son wanted to join the business, but nothing he sculpted had the precision or exactitude of his father's work. The boy asked his father, "How do you produce such beautiful work, while I seem to be completely inept?" The sculptor explained, "When *you* work, you try to carve the marble until it takes on the shape you desire. But I look inside the marble, and when I see the image that is trapped inside, I chip away at the blockages until it is revealed."

This parable can be explained as follows. The definition of a human being is his soul. The soul is hewn from beneath G-d's holy throne and is replete with spiritual beauty and splendor. Each and every one of us is made in the image of Hashem and we consequently have innumerable holy traits.

However, if that is the case, why don't people appear that way? Why do most people (ourselves included) seem to be base, materialistic, and self-centered? The answer is that our essence is covered over with all sorts of blockages. We have sunk into the un-G-dly world of self-service and committed sins that create thick, dense barriers between us and our Creator. If we would only clear away these obstructions, our beautiful souls would shine forth and cleave to Hashem in the most sublime fashion.

Now we can answer our earlier questions. It may be true that the *besulah* has yet to fulfill certain positive mitzvos. But the focus of Elul is not so much about building on our initial holiness as it is about restoring ourselves to our original pure level. Since Elul is the month of *teshuvah*, we must make the pure maiden our role model as we remind ourselves that we had a pristine beginning. Although we may not avoid life in our struggle to remain pure, we can always work to return to that initial purity with which we were all created.[1]

By now I think we can suggest what we should learn from Gad. Although it is true that we are enjoined to perfect ourselves, it is essential to recall that a human being already has some intrinsic perfection within his person. This is why Gad—the one who was born circumcised and "perfect"—is aligned with Elul. Gad reminds us that we too were born with many great qualities of perfection. He teaches us to be like that sculptor who kept his eye on the intrinsic perfection that always existed—even if (in our personal lives) it entails scraping away at all that covers it in order to allow it to come to light. While we must indeed search for our flaws, we must also recognize that all we need to do is scrape them away to reveal the G-dly image that has always been there, hovering beneath the surface.

To sum it up, *teshuvah* is often translated as "repentance." This English word has the connotation of feeling sorry, as it is etymologically related to the word "penitent." However, I think a more accurate translation would be "returning to the original and rightful place" (as in *hashavas aveidah*—returning a lost object to its original and rightful owner). When we say Elul is the month of *teshuvah*, we mean that it is the time to erase our sins so we can return to the original and rightful level at which we began. Somewhere beneath all the grime there is a *tzelem Elokim*, an image of G-d, and like the master sculptor in the parable, if we can keep our eye on the beautiful image within, we will be able to clean away the blockages until it is revealed.

---

1. What we have been writing here is, to a degree, based on the *Bnei Yissaschar* (*Ma'amar* 1 *L'Chodesh Elul, os* 6). He writes: *This month [of Elul] which is designated for* teshuvah, *has the* mazal *of* Besulah, *for through* teshuvah *one becomes like a new creation, like before the evil inclination imposed himself...*

A Detergent Called *Teshuvah*

Elul is the final month before Rosh Hashanah. As we know, each month has its own *avodah*—a distinct way in which we are supposed to serve Hashem. Let's see if we can pull out the lessons of Elul and thereby become better and holier people.

We know that *Besulah*—Virgo, the Virgin—is Elul's constellation and should therefore be our role model for the *avodah* of the month. Ostensibly, it should be very easy to make the *besulah* our role model. A virgin is a symbol of purity, as she is chaste and untouched. Similarly, one should make it his goal in life to be pure and free of sin.

On the other hand, one could make a fair argument that she is *not* a good role model for us. While purity has a time and stage when it is appropriate, Judaism doesn't idealize remaining a *besulah* throughout one's life. At some point one is supposed to move on and get married. There is a positive aspect to being a *besulah*, but isn't there also a positive benefit of being a wife and mother? If this is the case, how can we learn from the *Besulah*?

Before attempting to answer this question, let's add another: In his analysis of Elul, the *Bnei Yissaschar* (*Ma'amar* 1 *L'Chodesh Elul, os* 2) notes that the words "*mazal Elul Besulah*" have the same *gematria* (numerical value) as "*eim habanim semeichah Hallellukah*" [that is, the Hebrew words that mean "Virgo is the constellation aligned with the month of Elul" have the same numerical value as the Hebrew words from *Psalms* 113:9 that mean, "A joyous mother of children, praised be G-d!"]. The *Bnei Yissaschar's* discussion of this fact is deeply Kabbalistic, and we will not delve into it. However, one should still be astonished by this *gematria*. After all, we know that *gematrios* indicate equivalence between the words or phrases that share the same numerical value. But how can a *besulah* be equated with a joyous mother of children? Is this not a physical impossibility?

In order to answer these questions, we will have to take a detour to analyze the well-known *avodah* of Elul—*teshuvah* (repentance). Our Sages teach that the forty days between Rosh Chodesh Elul and Yom Kippur are days when Hashem is especially gracious and accepting of our *teshuvah*. We find this in the Torah when Moshe pleaded with Hashem for mercy for the Jewish nation—which was culpable for having worshipped the Golden Calf. The appeasement process began on the first of Elul, and full forgiveness was finally granted on Yom Kippur when Hashem stated, *"I have forgiven because of your words"* (*Bamidbar* 14:20). But how is *teshuvah* accomplished? What must we do to achieve forgiveness for our wrongdoings?

The *Sha'arei Teshuvah* writes (*sha'ar* 1, par. 19) that although there are twenty different aspects to a full and complete *teshuvah*, there are three ingredients that are absolutely essential: 1. *Vidui*—recalling something we did wrong and admitting it [mentally, verbally, in written format, etc.]. 2. *Charatah*—regretting and feeling pained over having sinned. 3. *Azivas Hachet*—making plans for how to cease committing the sin—now and in the future.

Now, the first and third ingredients are very logical. When one wrongs his friend, he admits that he did something wrong and then accepts to never do it again. But what is the explanation of the middle step—*charatah*? Why is it so essential for us to be pained over our past deeds? Why does Hashem require this in order to get back into His good graces?

The *Sha'arei Teshuvah* provides the answer in a different part of this chapter (par. 13). He writes that *charatah* is an essential ingredient because it's the "detergent" that washes away the sin. If we would only admit that what we did was wrong and accept not to act wrongly in the future, we would still be stained by the past action. But *charatah* is the tool that Hashem provides through which we can wash the sin from our record. *Sha'arei Teshuvah* writes: *The sin is lightened in direct proportion to the amount of regret one feels over having committed it.* Is it not a great kindness that Hashem allows us to wipe away the sins we have committed?

Now we can backtrack and answer our questions. Of course, it is not proper to avoid life in order to maintain one's purity. One must occupy himself with life and interact with others in order to accomplish the most he can. It is far better to engage in life and become a joyous mother of children than it is to avoid life and retain the purity of a virgin. If, because of the holy work, one becomes sullied, the person must ostensibly accept this as an unfortunate consequence. But what if it were possible to retain the positive aspects of both the joyous mother and of the *besulah*? What if one could engage in life, accomplish great things, get sullied as one inevitably does—and then clean away the filth and return to the initial stage of purity, with all accomplishments in hand? Wouldn't that be ideal?

Well, my friends, that is the secret of *teshuvah*. *Teshuvah* is the tool through which we can wipe away all the wrongdoings that come from our everyday lives. We don't *have* to avoid a life of accomplishments to be pure. Through *teshuvah* it's possible to be both a *besulah* and a joyous mother of children.

May we all merit utilizing the gift of *teshuvah* that Hashem gives us, and may we grow to accomplish great things while simultaneously retaining the purity with which we began.

Wakeful Warriors

It is said that the yeshivah students in Europe would tremble when Elul approached (and I don't think it was because summer vacation was ending). Elul is a serious time during which it is incumbent on every responsible Jew to prepare for the upcoming Days of Awe. But what is the growth of Elul? In which particular way are we supposed to improve?

The only calendric observance associated with Elul is the daily shofar blowing. The Rambam (*Mishneh Torah, Hilchos Teshuvah* 3:4) explains the purpose of the shofar: *Wake up you slumberers from your sleep!* (i.e. the purpose of the shofar is to wake us up). But what does the Rambam mean? Why is the metaphor of sleeping chosen? Wouldn't it be more logical to say that the shofar is meant to inspire us to repent from our misdeeds?

Let's ask another question. The *Sefer Yetzirah* says that the *avodah*, specific job, of Elul is to "rectify our deeds through wisdom" (*l'saken hama'aseh al yedei chachmah*). What does this mean? How can we accomplish this directive?

In order to answer these questions, let's take a detour: Each month is associated with a particular tribe, with Elul being matched up with Gad. In the *parshah* of *V'zos Habrachah* (33:20) we find a fascinating description of the unique qualities of this tribe: *Of Gad he said: "Blessed is He who broadens Gad; he dwells like a lion, tearing off [the] arm and even [the] head."* In this verse, Moshe is describing the great military prowess of the Gadite warriors. According to the simple understanding, Moshe is comparing the Gadites to a lion and describing how they would strike in battle with such ferocity that not only would the enemy's arm be cut off, but his head would also be shorn off in the same blow!

Now, why couldn't Moshe simply say that the Gadites were fierce warriors? Why does the Torah, which does not include a single extra word, need to inform us of this particular gruesome detail? The following explanation appears in the *sefer Shivtei Yisrael* (*Ma'amar* 1 *L'Shevet Gad*): The Torah is describing the details of the battles in which Gad fought while simultaneously alluding to another great quality of this tribe. The arm represents action while the head represents thought. *Shivtei Yisrael* explains that the single greatest fault that causes people to make mistakes is when there is a "disconnect" between their thoughts and their deeds. Moshe was praising the Gadites for how they were able to masterfully keep the two connected. If they would think of something that needed to be done, they would follow through and put the thought into action. And if they did something, it was only after thinking it through beforehand.

Now let's turn back to Elul. The great Rav Yisrael Salanter explains the "sleeping" that the Rambam referred to by pointing out that man is a creature of habit. He does what he does because he is accustomed to doing so. This great problem is what the *ba'alei mussar* refer to as *mitzvas anashim milumadah*—mitzvos performed by rote. Unfortunately, most people go through the waking part of their lives in a state of sleep. When a person is asleep, his brain is shut off. His body may be functioning, but it is without any cognitive participation on his part.

It is a terrible thing to live with one's mind disconnected from his deeds. One issue it engenders is that he will act without thinking about whether his deeds are proper or not. Another issue is that he may take the time to think about the proper thing to do, but his idealistic thoughts won't translate into action. Either way, the issue is that his head is not connected to his arm and he is, in essence, asleep.

Now we can understand our job during the month of Elul. The *avodah* of Elul is to wake up. We need to become like Gad and connect our brains to our arms. The *Sefer Yetzirah* writes that our goal is to rectify our deeds through wisdom. This is not as abstruse as it sounds. It merely means that we must start thinking before we act. The action will be "rectified" (i.e. done right) if it is thought out beforehand.

But if we are creatures of habit, how can we accomplish this? The answer is, through the shofar. Every morning we do a positive action that serves as an alarm clock to wake us up and shake us out of our mindless waking stupor. When we hear the shofar blast we are supposed to remind ourselves to think through our daily routines and decide whether or not they are proper. We do it again, day after day, until hopefully by the time Rosh Hashanah comes around, we are a little more cognizant in our daily routines of what we are doing and whether it is proper or not.

May we merit growing in this area and in becoming bigger and better people.

Tishrei

Angels, Exiles, and Cotton Balls

Tishrei is the most chock-full month of the year.[1] There are so many holidays packed into these thirty days that it would be arrogant to try to sum it all up in these few pages. Let's focus on Sukkos and see if we can pull out an encompassing theme and perhaps also connect it to the Yamim Nora'im, the Days of Awe.

Why do we build a sukkah and live in it for seven days? The *Shulchan Aruch* (*O.C.* 625:1) states: *The* pasuk *commands,* You shall dwell in *sukkos* for seven days, for I caused you to dwell in *sukkos* when I took you out of Egypt. *The* sukkos *under discussion are the Clouds of Glory that Hashem surrounded us with to protect us from the heat and wild beasts of the desert.*

The *Shulchan Aruch* is aiming to explain that the halachah concurs with the opinion that the *sukkos* in the verse are referring to the Clouds of Glory, as opposed to the opinion that they refer to actual booths built by the Bnei Yisrael. The relevance of this information is that we must actively think of those Clouds as we dwell in the sukkah if we wish to fulfill the mitzvah properly (*Mishnah Berurah, O.C.* 625:1). However, one could ask a fair question. I would understand if Hashem commands us to build booths as a remembrance of booths. But why must we build booths as a remembrance of clouds? Why can't we just stay in our homes and cover the walls with cotton balls? Why do we need to leave the house?

Perhaps we can answer this question with an idea from the *Pesikta D'Rav Kahana* (29). On Rosh Hashanah, Hashem passes judgment over the whole world, and on Yom Kippur He seals the verdict. Since we may have been found lacking, we leave our

1. The fact that Tishrei has so many holidays makes it difficult to explain the proper implications of its letter, tribe, or *mazal.* The following essays will rather focus on the mitzvos of Tishrei.

homes shortly after Yom Kippur to fulfill any decree of exile that may have been made against us. (This idea would also explain the famous question of why Sukkos is placed in Tishrei instead of in Nissan when the Bnei Yisrael left Egypt and were given the Clouds for the first time.)

But even this answer is incomplete. Even if we allow for Sukkos to simultaneously be a recall of the Clouds of Glory and also a fulfillment of a decree of exile (a difficult idea to swallow altogether), we would still have to wonder why Sukkos is called *Zman Simchaseinu*—the Time of Our Rejoicing. While it is certainly an accomplishment to relieve ourselves of any negative decrees, it seems inappropriate to refer to a time of exile with a joyous nomenclature.

Before we try to answer these questions, let's ask a third. Why do we take the *arba'ah minim*, the Four Species, on Sukkos? The midrash (*Vayikra Rabbah* 30:2) explains that the *arba'ah minim* are like a victory banner held aloft and waved by triumphant soldiers returning from the battlefield. We too wave the victory banner as we emerge from the Day of Judgment. This midrash is baffling when viewed in conjunction with the last point we made. Do we walk away from the Yamim Nora'im worried that we were found lacking and in need of exile, or do we walk away confident in our victory? And to make matters more confusing, the Arizal writes (see *Mishnah Berurah* 651:34) that it is best to take the Four Species (the victory banner) inside the sukkah (the exile symbol)! Can there be a greater contradiction than that?

To begin answering our questions, let's turn to a quote the *Bnei Yissaschar* brings from the *sifrei Kabbalah*. Explaining the aforementioned midrash, the *Bnei Yissaschar* (*Ma'amar* 10 *L'Chodesh Tishrei, os* 18) asks why these four species were chosen to comprise our victory banner. He explains that they are the exception to a very broad rule regarding Divine Providence that every creation in the world is maintained by its own personal angel. Obviously, the power comes only from Hashem, Who never loses sight of a single creation, and the angels have absolutely no power of their own. But the flow of blessing is usually sent through an intermediate angel. The Four Species, on the other hand, are maintained by G-d Himself. Therefore, the *Bnei Yissaschar* explains, they are perfect choices to form our victory banner—as they are from the "house of the King."

Now let's check out a similar idea from the Vilna Gaon (*Aderes Eliyahu, Ki Sisa*) regarding the Clouds of Glory. In the aftermath of the Golden Calf, Hashem declared, "I shall annihilate them!" Moshe prayed until finally Hashem relented. But there were consequences. Until now Hashem Himself accompanied His children. Now Hashem said, "I will send an angel before you." Moshe objected, "If Your Presence does not go along, do not bring us forward! [Rather, what I desire is that] Your people be distinct from all the people on the face of the earth!" Finally, Hashem gave in and stated, "Even this I shall do."

The Gaon explains that the Clouds of Glory came to fill the dual request for G-d's Presence and His distinctive care. We see from here that not only do inanimate creations have either direct Providence or an intermediate angel, but nations do as well. We also see that Hashem granted us—the Jewish people—a direct Providence without any intervening angel.

Now we can answer all our questions. We asked why we leave our homes. I believe the answer is that it would not be sufficient to hang cotton balls inside to recall the Clouds of Glory because the mitzvah is to relive and imbibe the reality of Divine Providence. While obviously our houses are also completely in Hashem's control, staying inside would lend us Hashem's protection through an intermediary, i.e. the roof. Only under the sky can we truly live the message of Hashem's direct guardianship.

As for our question regarding the contradiction of the sukkah being a joyous abode and simultaneously an exile, perhaps we can now explain that any other nation must indeed worry when forced to leave their land. If their patron angel has fallen out of favor, then what hope do they truly have? But Jews are different. Even in exile we enjoy Divine Providence, as Hashem joins us there. As it says: *And despite all this, while they are in the land of their enemies, I will not be revolted by them nor will I reject them* (*Vayikra* 26:44). So while it may be true that we go to the sukkah in order to be in exile, we can still simultaneously recall the Clouds of Providence that accompany us into the Diaspora. And for this we rejoice.

To sum it up, let's answer our third question. Do we walk away from Yamim Nora'im confident or worried? The answer is—both. We worry that perhaps we must go into exile. But we can take the *arba'ah minim* into the sukkah because we know that just as they have no intervening angel, we too are under Hashem's direct Authority. Even in the sukkah, even in the exile or the desert, He is our Sole Caregiver. We are truly in good Hands.

The King, the *Kapparah*, and the Cloud

The holiday that takes place within a month is usually our guiding light for what we can learn from that month, but Tishrei has so many holidays that it's hard to know where to start. Let's analyze the individual themes of the Yamim Nora'im (the Days of Awe—Rosh Hashanah and Yom Kippur) and then of Sukkos and see if we can put them together to draw an enduring message from which we can grow.

Everyone knows the theme of the Yamim Nora'im. Rosh Hashanah is the day of *Malchuyos*—when we make G-d our King. Yom Kippur is the Day of Atonement. The days in between are the Days of Repentance. Together, these days are called the *Yemei Din*—the Days of Judgment. The *machzor* states: *On Rosh Hashanah [the decree] is written and on Yom Kippur it is sealed.* Thus, we see that the first third of Tishrei is the time during which Hashem analyzes our deeds and uses them as the basis for the judgment of how the next year will look.

The theme of Sukkos, on the other hand, is a little trickier to pinpoint. Let's turn to the *Shulchan Aruch* (O.C. 625:1) as a reference point: *[The Torah commands,] You shall dwell in booths (*sukkos*) for seven days because I caused the Jewish People to dwell in booths (*sukkos*) when I took them from the land of Egypt. These booths (*sukkos*) mentioned in the verse are actually a reference to the Clouds of Glory.*

It seems from the *Shulchan Aruch* that the holiday of Sukkos is all about remembering the Clouds of Glory. Indeed, the *Mishnah Berurah* states: *When one sits in the sukkah, he must have* kavanah *(active intent) and think about the Exodus and the fact that Hashem surrounded us with the Clouds of Glory to protect us from the elements. This* kavanah *is necessary in order to fulfill the mitzvah of sukkah properly.*

What is it about these Clouds that we want to recall? It seems that Hashem surrounded us with them in the wilderness to protect us from the desert sun and the sweltering heat. Scorpions and snakes could not penetrate these Clouds, nor could enemy armies. It could then be said that the Clouds of Glory were Hashem's way of enveloping us in His all-powerful and loving embrace.

However, this idea is difficult. In what way does the sukkah remind us of the Clouds of Glory? If you were to say that Hashem took care of us following the Exodus and protected us from the elements when we were living in actual booths (as one Sage in the Gemara indeed opines), then I would understand why we should live in booths as a remembrance. But if, as the *Shulchan Aruch* says, the goal is to remember the Clouds of Glory, then how is that accomplished by living in a booth?

And let's ask a few more questions about this booth. What is the meaning behind the halachah that a sukkah must be temporary (i.e. shorter than a height deemed to be permanent)? Why did Hashem command that while the walls can be strong as steel, the roof must be made of weak natural growth? Why must the roof be of a thickness that results in more shade than sun, while at the same time the stars must be able to be seen? What is the meaning behind all these laws and how do they connect to the Clouds of Glory?

In order to answer all these questions, let's turn to the *arba'ah minim*—the Four Species that Hashem commanded us to take and wave on Sukkos. The midrash (*Vayikra Rabbah* 30:2) compares the Four Species to a victory banner held aloft and waved by triumphant soldiers returning from the battlefield. Why these particular species were chosen to be part of this victory banner is a discussion beyond the scope of this essay, but the idea of the midrash remains. And why are we celebrating? The midrash explains that it's because Yom Kippur just ended and the judgment was sealed, so we rejoice with confidence that we were acquitted of our sins and judged favorably.

But herein lies an essential question. What makes us so sure that Hashem decided we were going to have a sweet year? We know that every year has problems, so why are we waving a victory banner so joyously?

The answer is that it's true that we don't know what our decree contained. And it's true that it may contain hardships or bitterness. But it is coming from Hashem and therefore it is, by definition, good. Hashem does everything for a good reason. Everything He does is for our benefit. True, it is difficult to see Hashem in this world. His light is often not visible. Most of the time there is more shade than sun. But we are enjoined to recall that this world is but a temporary place. Hashem is above time and space; He is infinite. He knows better; He's got a plan. We can work as hard as we want, we can build skyscrapers of steel, but at the end of the day we are under the dominion and control of the Creator of the Heavens and the Earth. He displayed His omnipotence in the wilderness when He surrounded us with mere Clouds and they somehow sufficed to protect us from fierce beasts and marauding armies. His Hand may not be as visible and obvious anymore, but in spite of the fact that there is more shade than sun, the stars can be seen twinkling through the cracks. Hashem is indeed always *meitzitz min hacharakim*—peering through the lattices (*Shir Hashirim* 2:9). He's always watching us and everything He does is good.

Now I think we have what we can learn from Tishrei. On Rosh Hashanah, we declare Hashem King. We accept His yoke and beg for a sweet new year. We have ten Days of Repentance in which we try to show Hashem that we indeed wish to come closer to Him. On Yom Kippur, we beg forgiveness for all our misdeeds and plead that Hashem seal us in the Book of Life. Then, on Sukkos, we celebrate. We live in the sukkah and wave our victory banner. Why? Because the sukkah, with all its minute details, displays how we will try to live according to the reality we perceived during the Days of Awe: that Hashem is the omnipotent King of the universe. We will accept Him and all He decrees for us. Not resentfully—au contraire, joyfully! We know that He knows best and He only wants what's best for us. He has a master plan, and we have the privilege to watch it unfold.

Can you help but feel the joy?

Booths and Branches

It really shouldn't come as a surprise. Doesn't the calendar say that Sukkos is four days after Yom Kippur? So why do we go through the same routine every year? We finish *davening* on Yom Kippur, we go home to break our fast, and halfway through the meal somebody exclaims in shock, "*It's almost Sukkos!*" and we thus begin the annual mad dash to build a sukkah and buy the *arba'ah minim* (Four Species). Comical almost, but in truth it's understandable. The Yamim Nora'im, Days of Awe, are awesome (literally), and they consume all of our focus—leaving no room to think about any future events, even holidays. But every year, with Hashem's help (and no small amount of effort on our part), it usually all falls into place and we manage to walk into the Festival of Booths with a sukkah built and a beautiful set of *arba'ah minim* in hand.

However, there is one thing that usually gets the short end of the stick. Namely, how much do we study the mitzvos of the holiday? When it comes to most Yamim Tovim, we have time to prepare ourselves by breaking out the books for a period of time beforehand. We start learning the laws of Pesach thirty days beforehand. People usually start thinking about Purim on Rosh Chodesh Adar. Shavuos is counted down to (or rather, up to) for forty-nine days in advance. However, due to the tremendous focus we have on the Yom Hadin throughout the duration of Elul, we often don't get to learn enough about

Sukkos to sufficiently understand the holiday and the mitzvos it contains.

This is no small problem. Should we be celebrating Sukkos (or any Yom Tov) with no more understanding than we did when we first learned about it in elementary school? Rav Yaakov Kamenetsky used to tell the following parable to describe the tragedy of celebrating Yamim Tovim with nothing but a juvenile level of understanding: A young boy (let's call him Moishy) turns three and gets his first haircut. His parents buy him his first *kippah* and *tzitzis*. The *tzitzis* are adorable, just the right size for a three-year-old, with a colorful choo-choo train painted on them. He looks really cute. Moishy grows, and when he turns five, the *tzitzis* are a bit too tight on him. When Moishy turns seven and the *tzitzis* are all but strangling him, he looks like a clown if he is still wearing them. And if our young man insists on wearing his choo-choo train *tzitzis* to his bar mitzvah, he will look so patently ridiculous that we might just question his mental capacity! Onlookers would say, "Doesn't he realize that he is bigger now? He has outgrown those childish *tzitzis*. It is time for him to observe the mitzvah in a way befitting an adult."

Rav Yaakov used to say that this also applies to our celebration of the holidays. We must not view them with nothing but the childish understanding we picked up in school. The Torah is infinitely deep, and by extension the Yamim Tovim are as well. Let's start this journey together and see if we can't plumb some of the infinite depths of this holiday and maybe grow a little in our understanding of Sukkos and its mitzvos.

The first step in understanding the mitzvos of Sukkos is to discover what exactly these mitzvos are. The midrash (*Vayikra Rabbah* 30:2) states that there are seven mitzvos on Sukkos: the first four are to take the *arba'ah minim* (*lulav*, *esrog*, *hadassim*, and *aravos*); the others are to live in a sukkah, to sacrifice the *Korban Chagigah*, and *simchah*—to rejoice. Since we don't have the Beis Hamikdash, we cannot bring the *Korban Chagigah*, so let's focus on the other mitzvos—the sukkah, the *arba'ah minim*, and *simchah*.

A question immediately jumps out: What is the connection between these mitzvos? Why do we live in a sukkah for a week and what does that have to do with the Four Species? And what's the connection between these two mitzvos and *simchah*? Does each mitzvah have its own separate reason and they coincidentally all fall out on the same holiday, or is there perhaps some way in which they are connected? We Jews don't believe in coincidence, so they must be related in some way—but how?

Let's take a hint from the Torah. Regarding Sukkos, the verse in *Vayikra* (23:39) reads: *...when you bring in the crop of the Land, you shall celebrate Hashem's festival...* It seems that Sukkos is related to the gathering in of the crops. Indeed, Sukkos is also called *Chag Ha'asif*—the Festival of the Gathering, because this time of year is the end of the agricultural cycle. After a year of plowing, planting, watering, harvesting, and drying, the crop is finally ready to be gathered in and used. Finishing a successful cycle is definitely a good reason to celebrate. This might sound distant to all of us who are not farmers, but since we all eat food, we are inextricably dependent on the agricultural cycle. Therefore, we surely should celebrate when one such cycle is completed successfully.

Our Sages teach that the Four Species are actually a celebration of the completion of the agricultural cycle. We take four perfect representatives of all different members of the Plantae Kingdom and use them to thank Hashem for sustaining us for another year. We also find the idea that the *arba'ah minim* are a celebration in a spiritual sense. When Yom Kippur ends and the judgment is sealed, we rejoice with confidence that we were acquitted of our sins and judged favorably.

So how does this understanding of the *arba'ah minim* as a celebration connect to the mitzvah of sukkah? Well, *simchah* and rejoicing are good and well, but too much can be dangerous. It is good to celebrate the gathering of the crops, but when someone fills up his house with an abundance of blessing, this opens the door to arrogance. The Sages teach that one only rebels against Hashem amidst satiety, as it says, *And Yeshurun grew fat and he kicked* (*Devarim* 32:15). Confidence in a favorable judgment is commendable, but there's a fine line between confidence and over-confidence. So, particularly at this juncture, when the cupboards are full and we risk becoming overly comfortable, Hashem sends us out of our houses into simple huts. "Go into a temporary dwelling," says Hashem. "Live under the stars and remember Me. *I* am the One Who takes care of you. It always has been Me, since the days of the desert when I surrounded you with nothing but Clouds and yet you were safe. Just as it was Me then, it is Me now as well." Hashem also tells us, "Celebrate and rejoice in the goodness that I gave you. But don't get carried away with your physicality. You are only here temporarily. Remain humble."

Thus, we see that the *arba'ah minim* and the sukkah balance each other out. The *arba'ah minim* raise us up and the sukkah keeps us grounded. Perhaps that is also the reason we read *Koheles* on Sukkos. Hashem wishes to remind us that while we rejoice in our ingathering of the crop, we should remember the temporal nature of This World and not get carried away. Can anyone possibly become arrogant about his mastery over a world that is wholly futile? It would be difficult.

Rav Samson Raphael Hirsch also understands the *arba'ah minim* and the sukkah as two mitzvos meant to play off each other, but in a bit of a different manner. He writes: *The sukkah [in its simplicity] negates material possessions as ultimate value; the* lulav *(read: the Four Species) teaches us to…dedicate our possessions to the sacred [service] of G-d… From this wells the fountain…of happiness… for he can fulfill the will of G-d if he has much or little* (*Horeb*, ch. 31). Rav Hirsch is saying that the sukkah teaches us to keep our physicality at bay, and the *arba'ah minim* tell us that we must also use the physical to elevate ourselves and the world. (Additionally, he adds that this will lead to true *simchah.*) In this way, we find a second connection between the mitzvos of the holiday. Let's keep going!

This explanation of Rav Hirsch calls to mind a similar explanation the *Sefer Hachinuch* gives for the mitzvah of *arba'ah minim*. The *Chinuch* (mitzvah 285) explains that Hashem wants to give us as many merits as possible. To that end, He commanded us at the time of the gathering to turn the best of our crop into a mitzvah that causes us to think of Hashem and praise His

Name. We see from here that it is very important to always keep our focus on Hashem, even in seemingly mundane matters. We take our representation of the epitome of physicality (the most beautiful representatives of all different types of crops) and we bring them into the sukkah (i.e. under the Heavens). Thus, we form a harmonious blend in which we rejoice for all that we have, while simultaneously acknowledging from whence it came. In this approach as well we find that the mitzvah of sukkah is meant to work together with the *arba'ah minim* to help create an appropriate and balanced relationship with our Creator.

In my (hopefully humble) opinion, I think we can take this idea of Rav Hirsch and the *Chinuch* a step further. You see, Sukkos takes place immediately following Yom Kippur. On Yom Kippur we try to be like angels. We don't partake of food and drink, we stay in shul all day, and we try to come closer to G-d by doing so. But at the end of the day, we are not angels. We are physical beings with physical needs. Maybe Yom Kippur represents the ideal. But we need to eat and drink, and we cannot live in shul all year round. A house is a necessity. So are we doomed to forever be constrained by our human limitations? No. Hashem let us in on the secret. The physical doesn't have to be an impediment; au contraire, it can be utilized! If we keep our dwelling here on an appropriately simple level and we are wary not to overly indulge, we can merit to dwell in the *shade of the Almighty*. We can use the physical fruits and branches to serve Hashem and grow closer to Him. In this light, the Four Species and the sukkah provide the same message—namely, the secret of how practically to become spiritual while locked in a physical world.

Thus far we have suggested three slightly different ways of explaining how the sukkah and the *arba'ah minim* are opposites that balance each other out, and we described one way in which they deliver the same message. Let's continue; there are still more connections to find.

There is a famous midrash (*Vayikra Rabbah* 30:12) that describes how the *arba'ah minim*, in their diversity, represent all different types of Jews: *Just as the* esrog *has taste and aroma, so too there are*

Jews that have Torah and good deeds. Just as the date palm produces [food that has] taste but the palm has no smell, so too there are Jews with Torah but no good deeds. etc. The midrash continues: *Bind them all together…* We see a beautiful concept from this midrash. *Achdus*, unity, is a core value of the Jewish people. To be inclusive and caring for all of our own is ingrained in our spiritual genetics. This is represented in the inclusion of the malodorous *chelbinah* spice in the *Ketores* incense of the Beis Hamikdash and in the fact that the Hebrew word *tzibbur* (congregation) stands for ***tz****addikim* (righteous ones), ***b****einonim* (in-between ones), ***u'r****eshaim* (and wicked ones). It would seem then that one of the messages of the season is *achdus*. But isn't *achdus* a value all year? Why do we have this mitzvah specifically at this time of the year, and what does the sukkah have to do with it?

I think the answer can be found in the *Kli Yakar* (*Acharei Mos* 16:30) that explains why Moshe gathered everyone immediately after Yom Kippur to start building the Mishkan: *It is normal for people to beg for forgiveness on Yom Kippur and then return to their evil ways the very next day. Therefore, Moshe gathered them on the day following Yom Kippur, for it was then that they were united as one group. This is so because all fights are over one's involvements [in the world]…*

The *Kli Yakar* is saying that throughout the year we are steeped in our worldly pursuits. Ever since Adam was cursed to eat by the sweat of his brow, Man has had to work for a living. Unfortunately, this often causes disagreements and fights, and all too often strife and hatred. On Yom Kippur we rise above all that. We ask each other for forgiveness and make up. Indeed, the *Mussaf Shemoneh Esrei* on Yom Kippur labels the day: *a day of instituting love and friendship, a day of forsaking jealousy and competition.* We fast all day, removing ourselves as far as possible from physicality as we strive for the Divine. We try to gain a new perspective on this world and hopefully reach a conclusion that we shouldn't overindulge; we can survive on much less.

The sukkah is our attempt to hold on to that momentum. If you realized that you can live a less-extravagant lifestyle on Yom Kippur, implement this new philosophy immediately by living in a simple hut (at least for a week). If you see that you really can get along with your brethren, whether they have Torah and good deeds or not, then maintain that momentum and stop using your possessions to raise yourself over others. Maintain this newfound unity and abandon your house in favor of a simple hut where you and everyone else dwell equally under the heavens. Only with *achdus* can the Divine Presence dwell, both in the ancient times of the Mishkan and in present times as well.

In a similar vein, let's see what the Vilna Gaon (*Gra, Shir Hashirim* 1:4) says on the placement of Sukkos in the month of Tishrei: *Why is Sukkos celebrated in Tishrei if the Exodus was in Nissan? The answer is because [the sukkah is a representation of the Clouds of Glory that G-d surrounded us with in the wilderness. So Tishrei is the appropriate time to commemorate that because] it was at that time that the Clouds returned, since they had left by the sin of the Golden Calf.*

Think about it. In whose merit did we have the Clouds of Glory? The Torah teaches that they came in the merit of Aharon—the *rodef shalom* (pursuer of peace). Does this not indicate how Sukkos is the season to work on brotherhood and peace? Both the *arba'ah minim* and the sukkah point to this message of *achdus*. Let's listen when they speak.

There is yet another famous midrash that might help us find a sixth explanation regarding the connection between the booth and the branches (*Vayikra Rabbah* 30:14): *All my body shall call out, "G-d, who is like You!" The spine is the* lulav. *The* hadassim *are the eyes. The* aravah *is the mouth. The* esrog *is the heart.*

This midrash is saying that a person is obligated to take all that G-d has granted him and use it in *avodas Hashem* (service of G-d). The main parts of the body and, by extension, the rest of the body as well, must sing out to Hashem in all situations, and an individual must always remember that his body was only granted to him for the purpose of serving the Divine. The sukkah contains this message as well. In one's house (or anywhere, for that matter), mitzvos can only be done with a portion of the body at one time. The laying of *tefillin* and the lighting of Shabbos candles involve only the hands, while *davening* involves the heart and mouth. Shofar is a mitzvah of the ears and going to visit the sick is a mitzvah of the feet. For the mitzvah of sukkah, however, one can use his entire body. Every limb, sinew, and blood vessel fulfills the commandment of the Creator. We completely leave our houses and go live under the "shade of G-d" for a week. He is the King of the world Who displayed His might in the desert and we accepted and re-accept His servitude upon ourselves. We bring our entire bodies under the stars and tell Hashem that we are completely subservient to Him. Sukkos is when we wholly and unrestrainedly give ourselves over to the service of G-d. Both the branches and the booth relay this message. May we merit to internalize it.

Let's move on with a seventh connection between the booths and branches. There is a somewhat less-famous explanation of the *arba'ah minim* given by Rav Samson Raphael Hirsch. Rav Hirsch describes how the Four Species parallel all that we have in nature. He compares the *hadas* (myrtle) with its sweet smell but lack of taste to fresh air or sunshine, which is pleasurable and refreshing but not directly sustaining. The *lulav*, whose tree produces food but no smell, is that which has sustaining abilities but is not readily accessible and rather must be extracted—like foodstuffs. The *esrog* completes the first and second category jointly, and the *aravah* (willow) is the area in which *man exercises his power...for he extracts all usefulness...through his own skill* (*Choreb*, ch. 31). A prime example of what the willow represents is wood. Wood does not have any directly sustaining abilities and it is also not readily accessible; rather, it must be extracted from its source (i.e. the tree). Wood is therefore an example of something that is largely in our (seeming) control.

So essentially, we have four categories of nature, and we are commanded to gather representatives from across all four categories and celebrate with them before Hashem. First, this is highly appropriate for the Time of the Gathering. But on a more profound level, we acknowledge that everything—from the sunshine over which we obviously have no control to the wood from which "we" form our "own" creations—is completely in Hashem's hands. As an interesting aside, Rav Hirsch thus explains why we beat the *aravos* on Hoshana Rabbah. He says: *...this [willow—that represents the creations of man—is beaten because this] denotes the compelling need to keep aloof from...the glorification of man's strength.*

The sukkah contains the same idea. It recalls the time when G-d protected us in a desolate, howling wilderness from the heat, the cold, the snakes, and the scorpions. It was blatantly obvious in that situation that Hashem was in complete control. We saw His omnipotence clearly. Logically then Hashem is still omnipotent, and He is still in complete control of the world. Together, the sukkah and the *arba'ah minim* provide a message that Hashem was, and still remains, the Master of the world. Is this not a reason to celebrate? We have an omnipotent, benevolent G-d on our side. The Clouds of Glory returned in times of old to show us that Hashem loves us and will always care for us—and that message still resounds to this day.

Celebrations and Cerebrations

The month of Tishrei is blessed with an abundance of Yamim Tovim—Rosh Hashanah, Yom Kippur, Sukkos, Hoshana Rabbah, Shemini Atzeres, and Simchas Torah. All of these holidays are important and deserve an in-depth analysis. Realistically, though, there is only so much one can fit into thirty days. So, more often than not, the latter holidays receive far less attention than the former. Perhaps it would be prudent to spend some time on the final one on the list—Simchas Torah.

What is this Yom Tov? We all know that Simchas Torah is the day on which we collectively complete the Five Books of the Torah. But why is this given its own holiday?

Obviously, one could simply (and correctly) answer that the mitzvah of Torah study is great because it equals all the other mitzvos combined (*Shabbos* 127a). However, this idea is enigmatic. *Why* is Torah study the greatest of all the mitzvos? Doesn't the Torah (*Pirkei Avos* 1:17) place a greater emphasis on deed than on study? It seems we need to delve further to understand this precious mitzvah.

To answer our questions, let's turn to the final verses of the Torah that consist of Hashem's eulogy for His most faithful servant: *Never again has there arisen in Israel a prophet like Moshe, whom Hashem had known face to face. All the signs and wonders that Hashem sent him to perform in the land of Egypt…the strong hand…the awesome power…before the eyes of all Israel.*

Since there are no extra words anywhere in the Torah, we are obligated to justify each one of these praises: *Whom Hashem had known face to face*—Rashi explains that this is referring to the fact that Moshe was "familiar" with Hashem in that he could speak with Him whenever he desired. *All the signs and wonders that Hashem sent him to perform in the land of Egypt*—this phrase is obviously referring to the Ten Plagues. *The strong hand*—Rashi elucidates that this is referring to the greatness inherent in Moshe's ability to bring the *Luchos* down from Heaven. Rashi's elucidation continues and explains that *the awesome power* alludes to the great miracles Moshe performed in the desert.

Thus far the eulogy seems to make sense. All these things were truly wondrous deeds that indicate that Moshe was a great man. But now let's view the final words of the Torah: *Before the eyes of all Israel.* Why does the Torah have to include these words? Isn't it obvious that the aforementioned wonders were performed before the eyes of Israel? Rashi explains that these words were written to allude to another time "the eyes of all of Israel" were mentioned. We see earlier in *Sefer Devarim* (9:17) that when Moshe recounted the sorry tale of the Golden Calf it says: *I grasped the two Tablets…and I smashed them* ***before [all] your eyes***. Thus, we see that in this eulogy, Hashem was actually praising Moshe for breaking the Tablets of the Law. Similarly, we find that Hashem praised Moshe immediately after the fact, coining a popular phrase, "Yasher koach *for breaking them, Moshe!*" (*Shabbos* 87a).

Now this is perplexing. Why are we ending the Torah on a sour note? Is it really necessary to make mention of one of the top five worst sins in the history of mankind right here and now? Don't we know the halachah that when we read the Torah publicly we are not supposed to divide the readings so that they end on a negative note? Granted, it was indeed the right thing for Moshe to do. But in what way was it such a special act that it warranted honorable mention in Hashem's concise and succinct eulogy of Moshe?

Let's ask ourselves—what exactly is the Torah? To the untrained eye, it may seem to simply be a compilation of stories about the early stages of the Jewish people. But one who studies the Written and Oral Torah in depth will (or at least should) quickly realize that there is a lot more to the Torah than just that. The Torah was written by Hashem Himself. It contains His views, His thoughts, and His approach to just about everything. This idea is extraordinarily significant. Anyone who is even somewhat acquainted with the Oral Torah (which explains every phrase, word and letter in the Written Torah) can tell you that the gamut of the Torah is so wide that there is nothing it does not include. The Torah teaches us how to think by teaching us precisely what Hashem's opinion is on every single matter under the sun.

This sets Torah apart from any other study. While the sciences, philosophies, and arts can indeed teach you to think and understand, they are limited in that they can only teach you to be as discerning as the one who formulated them in the first place. Torah, on the other hand, can teach you to think like the Almighty Himself—because Torah is His Mind (as it were)!

Consider the following astounding thought: What if I were to tell you that you can ask me any question on the Freudian approach and I could tell you with confidence what Mr. Sigmund would reply? Would you not be impressed? Or what if I would tell you that I could confidently express the views and thoughts of Albert Einstein? Would that not be an amazing feat? Now consider what you would say if I told you that I most assuredly can represent the views and opinions of the Infinite Creator of the Heavens and earth? Would you not be astounded with my brilliance? Yet that is precisely what many a Torah scholar can do. One who studies the Torah for many years and immerses himself in it completely and totally earns what we call *da'as Torah*. One who attains this lofty level is approached for advice by Jews around the globe. He is asked not only on matters of ritual, but even on everyday, worldly matters. The reason his advice is so precious is because we know he can accurately tell us what the Almighty knows is the best course for us to take.

Now we can understand why Hashem ended off the Torah alluding to Moshe's breaking of the *Luchos*. Perhaps the explanation is that Hashem could think of no greater praise than the idea that Moshe had reached the level where he could make a decision as impactful as he did without even consulting G-d and still divine the correct conclusion. In other words, Moshe reached a level where he had earned the greatest heights of *da'as Torah* possible. While the breaking of the Tablets might be a sour memory, Hashem ended off the Torah with the awesome and inspiring message that it is possible for a human being to reach a level where his mind is aligned with the Divine. Moshe did it, and through Torah study we all can do it as well.

Now we can suggest an explanation for why Torah study is the greatest mitzvah. While obviously fulfilling the Torah is immensely important, only through Torah study can we train ourselves to think like Hashem. This Yom Tov is all about man's incredible capacity to learn the Divine way of thinking and thereby connect to Hashem in a way no other mitzvah provides. This, my friends, is why we celebrate Simchas Torah.

Mar-Cheshvan

The Scorpion and the Setting of the "Son"

After a jam-packed Tishrei, the month of MarCheshvan presents an altogether different challenge: How do we grow *without* the guidance of the special holy days? Our journey thus begins—let's go.

Perhaps the most glaring aspect of this month is its name: MarCheshvan—Bitter Cheshvan.[1] Simply understood, it is a bitter time because there are no holidays to provide the sweet opportunity to come close to Hashem. However, this thought generates the disturbing suggestion that Cheshvan has no special qualities. Since that is obviously untenable, we are forced to conclude that since the month is bitter, its quality is bitterness! That being determined, we now have to discover how bitterness can be used in the service of Hashem.

Here's another question for the mix: It is well known that each of the Jewish months align with one of the constellations. The *mazal* of Cheshvan is *Akrav*—the Scorpion. Working to understand this can be a daunting task. All the Torah teachings about the calendar seem to work with a basic understanding that each month's characteristics can be used properly or improperly. However, no quality or trait is ever presumed to have an intrinsic malignance. Yet I put forth a challenge to my readers to find a single Torah source that views scorpions in a positive light. The scorpion is invariably a representation of a negative surprise that befalls the unfortunate. If this is the case, we should be asking ourselves what type of positive growth we can glean from this sinister arachnid.

---

1. Although the *Nachalas Shivah* and other sources indicate that the name of the month is properly MarCheshvan and not Cheshvan, the commentators still seem to understand that the part of the name which has the word "Mar" indicates that the month is bitter.

In order to answer these questions, let's take a detour to analyze the tribe aligned with Cheshvan—Menashe. The Torah does not discuss Menashe in great detail, but we'll have to work with what we've got. In *Parshas Vayechi* we find Yaakov Avinu at the end of his life. Yosef brings his two sons Ephraim and Menashe to receive their grandfather's blessing before he dies. Yosef positions his firstborn Menashe on Yaakov's right side so he would receive the greater blessing, which stems from the right hand. Then Yosef positioned the younger Ephraim on Yaakov's left so that Yaakov would bless him with his left hand. Yaakov proceeded to cross his hands over each other in order to bless Ephraim with his right hand for, as he said: *...the younger brother shall become greater, and his offspring's fame will fill the nations [and he is therefore more in need of the greater blessing]* (*Bereishis* 48:19).

Thus far everything makes sense. The difficulty lies in the wording of the *pasuk*: *...sikeil es yadav* ***ki*** *Menashe habechor*—"he crossed his hands '***ki***' Menashe was the firstborn" (ibid. 48:14). The Hebrew word *ki* generally translates as "because." However, in context, this makes no sense because Yaakov didn't put his right hand on Ephraim *because* Menashe was the firstborn, but rather *in spite of the fact* that Menashe was the firstborn. Many commentators therefore conclude that the word *ki* here must mean its highly uncommon translation of "in spite of." However, let's continue this analysis in accordance with those who take *ki* to mean its usual definition of "because" (*Chizkuni*, *Maharshal*). In that case, what does the Torah mean when it says Yaakov crossed his hands *because* Menashe was the firstborn?

I would like to suggest that perhaps the Torah is addressing an obvious yet unstated question. Why did Yaakov keep Menashe on his right if he thought it was more appropriate for Ephraim to be blessed with the right hand? He could have simply switched Ephraim and Menashe's positions rather than crossing his hands. The Torah answers: *He [acted strangely and] crossed his hands* ***because*** *Menashe was the firstborn.* Since Menashe was older, he still deserved the honor of being on the more important side. Ephraim, however, needed the greater blessing (which comes from the right hand), so Yaakov resolved the issue by crossing his hands.

It may sound like we answered our most recent question. However, there's still an issue. Let's ask ourselves, why did Yaakov think that Menashe should be kept on the right? Didn't he see that the younger son was more deserving? Apparently, Yaakov understood that the Torah requires us to give greater respect to the firstborn regardless of merit. However, if that is true, then what prerogative did Yaakov have to supersede Menashe's birthright by giving Ephraim the greater blessing? Perhaps you'll answer that Yaakov had a right to decide which son was worthier. But if that was the case, why didn't he place Ephraim on his right? Suggesting that Yaakov was simply compromising would almost imply that he was uncertain of what to do (unlikely). Philosophically speaking, is one permitted to choose according to his understanding or is one obligated to bow before the circumstances set in place by the Almighty?

The answer is that of course one is allowed—even obligated—to try and change any given situation (in accordance with what he perceives as the Divine Will). We are not permitted to put up our hands and absolve ourselves of responsibility, claiming that things are the way Hashem wants them to be. That being said, one's deeds may be very different if he realizes that in reality Hashem is in control and we only act because Hashem commands us to. One difference may be in the extent to which we try to alter a situation. A true believer will only change what needs to be changed but refrain from gratuitous modifications because he realizes that (barring the part that Hashem obligated him to change), the situation is exactly how Hashem wants it to be.[2] Worded differently, every situation that Hashem created is perfect with the exception of those situations that He created imperfect in order to allow us to perfect them. Therefore, we may only ever act where Hashem wants us to, but we must leave

2. Admittedly, it is usually quite difficult to ascertain the Will of Hashem regarding what is considered an appropriate change and what is considered gratuitous.

everything else as is. Perhaps this is why Yaakov was permitted to give Ephraim the greater blessing, but he refrained from moving Menashe. Giving Ephraim the greater blessing was necessary and therefore appropriate, while moving Menashe was unnecessary and therefore inappropriate.

Now perhaps we can backtrack and answer our questions. How can we grow from bitterness and what is the lesson of the scorpion? We need to recognize that there will always be times in our lives that are bitter, and sometimes bitter circumstances pop out and befall us. While we obviously must try our hardest to improve our situation, our deeds must be tempered with the realization that Hashem is orchestrating everything. Of course we must act, but if indeed we cannot improve our lot then we must calmly accept that some things are bitter and only Hashem knows what is truly good for us. If we can do this, our lives will be that much sweeter.

The Month of the Working Man

Cheshvan is the type of month that brings to mind the expression "twice the pride, double the fall." Tishrei is a month packed with holidays…and then comes Cheshvan, and what do we have? Nothing. It is the most mundane of all the months, with not a single Yom Tov. The only "happy" part of the month is that we finally have a chance to go back to work and earn a living! Not too exciting, is it? No wonder this month carries the name MarCheshvan—"Bitter Cheshvan." Yet we know that each month carries its own *avodah*, specific job. Hence, our goal is to discover in what way Cheshvan is special and how we should be serving Hashem during this time.

The Arizal writes that Cheshvan is aligned with the tribe of Menashe (since Levi and Yosef are generally not counted in the list of the tribes, they are replaced by Ephraim and Menashe). What do we know about Menashe? *Chazal* teach that while Ephraim stayed and learned Torah from Yaakov, Menashe helped

Yosef run the kingdom. It seems Menashe was a working man. Perhaps one could say, somewhat tongue in cheek, that that is why he is matched up with Cheshvan—the month when one is primarily at the office.

Now, of course, it is necessary to work. The world has to function, and *mann* doesn't fall from the sky anymore. But if each month is unique in the way we serve Hashem, then how can we explain the working month? As a great *rosh yeshivah* once told his student who was itching to leave yeshivah and join the workforce, "Don't forget, when Hashem told Adam, '*You shall eat by the sweat of your brow*,' it was a curse, not a mitzvah!" Yes, we unfortunately have to go to work. But wouldn't our *avodas Hashem*, service of G-d, be much greater if we could focus on spirituality all day?

One might suggest that Menashe's greatness was that he kept the kingdom going so Ephraim was able to learn. We see a similar arrangement regarding Yissachar and Zevulun; Zevulun would work and support Yissachar in his learning, with the result being that they shared the Heavenly reward. There is definitely truth to this idea. Indeed, *Pirkei Avos* (3:21) states: *If there's no bread, there's no Torah.* So perhaps the greatness of the working man is that he enables others to fulfill the Torah.

However, there still is a slight problem. Let's read the mishnah in its entirety: *If there is no bread there is no Torah. If there is no Torah there is no bread.* We elucidated the first half of the mishnah to be read: *If we didn't have bread, we wouldn't be able to have the Torah.* If we were to presume a parallel elucidation, the second statement would read: *If we didn't have the Torah, we wouldn't have bread.*

What does this mean? Granted, we need bread to eat and survive, and if we didn't eat and survive we would not be able to live and keep the Torah. But what does the mishnah mean when it says we need the Torah in order to have bread? Don't the nations of the world have bread even without having the Torah? (Maybe you'll suggest that the mishnah means that if we didn't have the *merit* of Torah, we would not be blessed with bread. But I think this suggestion should be discounted, for shouldn't we presume a parallel in the elucidation of the two halves of the mishnah? If the first statement said nothing of merits and connotes a direct connection, one ostensibly should assume that the second statement is not said in terms of merit either.)

Let's turn to the Ramchal to find our answer. In his great *mussar* work *Mesillas Yesharim*, the Ramchal writes: *All life situations in the world are tests for man. There is wealth on one side and poverty on the other. There is tranquility on one side and suffering on the other. Will one become rich and deny Hashem, or will he become poor and steal?* Simply understood, the Ramchal is teaching that man must struggle in each life setting to do the right thing and that each life setting provides a different challenge.

But perhaps we can understand this on an even deeper level. Let's turn to another one of the Ramchal's works, *Derech Hashem* (part 2, ch. 3): ***Man's task is to choose good...and not evil.*** *Mercy is a good trait, while callousness is not. Being happy with what one has is a good trait, while the opposite is bad... Man must choose from these good qualities and turn from the evil ones. In order for this to be possible, it is necessary for individuals to be put in different stations in life. Whichever station a person is placed in is a test for him, an opportunity to let him strive for the good and turn from the bad.* ***If there were no such thing as poverty or wealth, there would be no opportunity for people to choose generosity over indifference.*** *Will the rich man be haughty or humble? Wealth enables one to be tested in this aforesaid manner. The poor are likewise tested to see if they will be satisfied and thank Hashem for their lot* (emphases mine).

I think that now we can understand what is special about the workplace. Hashem created the world because He wanted to be able to reward us for choosing good. The angels wanted the Torah, but Moshe told them that they couldn't fulfill it. "Do you have parents to honor or bodies that pull you to sin?" he asked. The only way to fulfill the Torah is in the physical world. The challenges that the real world throws at us are not impediments to a life of spiritual growth; they are the *means* through which we live lives of spiritual growth. The workplace is where it's at! It is there that the Torah can truly be observed.

The *Zohar* (*Terumah* 161a) teaches: *Hashem looked into the Torah and created the world.* This means that it's not since people are born of parents that Hashem commanded us to honor them. Rather, since Hashem saw that it is good and holy to honor parents, He created a world in which people are born of parents—thereby enabling them to become holy by honoring those parents! With this *Zohar*, we can understand why without the Torah there would be no bread. The meaning of the mishnah is that if we didn't have the mitzvos of the Torah to fulfill, Hashem wouldn't have bothered to make the world in a way where we work to earn our bread. Meaning, the goal of the world and its life settings (poverty and wealth, etc.) is to provide the opportunity to become holy—like the Ramchal said. Hashem only created the world so that it could be the place in which we fulfill the Torah. If there was no Torah, the world would have been created without a need for bread.

Let's pull it all together with one final thought about Menashe: The Torah writes that Menashe was the royal interpreter (*Rashi, Bereishis* 42:23). What is the significance of this fact? I think we can suggest that Menashe was the one who understood what the king wanted and managed to translate the king's words to the common people. He conveyed the king's will to the more mundane areas of the kingdom and enabled it to be fulfilled. In MarCheshvan we need to become like Menashe in order to succeed in making Hashem proud.

Kislev

Eras and Archers

Once Kislev begins, people start to think about Chanukah. It is usually correct to assume that the holidays can help teach us the lessons we need to absorb during the month in which they are found. However, this concept may be difficult when it comes to Chanukah, as the holiday straddles two months. Let's take up the challenge and see if we can pull out a meaningful Kislev-related Chanukah lesson that can help us come closer to Hashem.

Chanukah is a difficult holiday to understand. We all know the story of the fierce war the *Chashmona'im* waged against the Syrian-Greeks in order to protect the Jewish way of life. We all know that following immense victories against all odds, the rededicated Temple Menorah burned miraculously for eight days. But the part that is strange is the fact that we make such a big deal over the miracle of the flames. Granted, it was a supernatural feat beyond the capabilities of any human being. But aren't we Jews already accustomed to miracles? We saw the Ten Plagues decimate Egypt. We saw the Red Sea split. We saw the heavens open up at Har Sinai and reveal the Creator Himself. We traveled with the Clouds of Glory, ate the *mann*, and watched the walls of Yericho fall at the sound of the shofar. We saw the sun and moon stand still at Givon and Ayalon. We saw ten daily miracles in the times of the first Beis Hamikdash. So why are we making a holiday over the fact that the flames did not burn out? What is so special about this relatively minor miracle?

In order to address this question, let's take a detour and ask another question. Where has G-d gone? Now, before you become alarmed by my question, let me explain. It is impossible to lose omnipotence. A human can gain power and lose power. But if one is all-powerful then it is absolutely impossible to lose any of that power. This is because weakening is something that only occurs as a result of some sort of deficiency. So where has G-d been for the past two thousand years? We have the unbroken transmission from our fathers that G-d indeed made all the above miracles and many more. We saw His omnipotence and we therefore know that He still is omnipotent. So, why haven't we seen any open miracles since the story of Chanukah? Does Hashem no longer care for His people?

This question can be answered by gaining an understanding of the eras and phases we've gone through as a nation. We used to be a nation of prophets. The Jewish people used to be on a very high spiritual level. We were learned in the ways of holiness and were able to access the world of spirituality. However, Hashem always wants us to be challenged in this world, so the ability to reach prophecy came along with the *yetzer hara*, evil inclination, for idolatry. Great and holy people succumbed to a temptation we can't even begin to understand—the desire to bow to idols.

Time passed and our level sank. The Sages saw that we could no longer stand up to the urge for idolatry so they prayed for that craving to be taken from us (*Yuma* 69b). They were successful, but the price we paid was that we no longer would be able to see Hashem with a clear revelation. No longer would the world of spirituality and prophecy be readily accessible to us. The world that presented the urge to serve idols would be distanced, at the expense of prophecy and open miracles.[1]

Now, perhaps we can answer our questions. Of course Hashem still cares for His people and can still perform miracles. But until the Messianic Era we remain in the Era of Concealment, in which Hashem hides behind nature. Now we can understand the greatness of the Chanukah miracle. It is true that we have seen many great miracles throughout our history and perhaps they were indeed more dramatic than the Chanukah miracle. But they weren't as big a deal as the Chanukah miracle because they took place during the Era of Miracles. By the time the

---

1. This is known as the concept of *zeh l'umas zeh* (lit. "this opposite this") (*Koheles* 7:4). The basic idea is that Hashem always keeps a balance in the spiritual world in which there are equal forces of purity and impurity in order to preserve our ability to have free will.

Chashmona'im rededicated the Temple, the curtain had been closed on prophecy and open miracles. The fact that there was an open and supernatural miracle during the Era of Concealment was definitely a big reason to celebrate.

Although we may have answered our questions, our work is still not done. Why did Hashem "break the rules" and make an open miracle during the Era of Concealment? What does Hashem want us to learn from this exceptional circumstance? The answer can be found in the Ramban at the end of *Parshas Bo*, where he discusses the purpose of open miracles: *[Hashem sometimes makes open miracles because] as a result of [seeing] the open miracles a person will come to [recognize and] admit to the hidden miracles...for a person must believe that all that occurs is miraculous. There is no [such thing as] nature!* The Ramban is saying that we are supposed to use open miracles as proof that Hashem is in complete control of the world.

I would like to suggest the following understanding. One might have thought that Hashem only watched over us back when we were worthy. But who says Hashem hasn't turned us over to the natural laws of the world now that we live in the Era of Concealment? To discredit this thought, Hashem made an exception to the rule and performed one single supernatural miracle shortly after the Era of Concealment began. Hashem "set the tone," so to speak, and informed us that even in this new era, He is in complete and total control. We are expected to use that one open miracle as a symbol by which we view every occurrence in our lives.

Now let's turn back to Kislev. The *mazal* (constellation) of Kislev is *Keshes*, which in English is called Sagittarius, the Archer. In what way is the Archer connected to the miracle of Chanukah and its attendant lessons?

Let's turn to the Gemara (*Bava Kamma* 22a) that discusses who is obligated to pay for damages. The halachah takes into account who owns the damager, who caused the damage, and how directly the damage was caused. The Gemara discusses the idea that if someone lights a fire, the flame may not merely be called "his money," but rather may also be viewed as "his arrows." The underlying logic of this is that in spite of the fact that fire travels and damages at a distance from the place it was released (as does an arrow), the action was still engendered by the one who lit it (as is the shooting of the arrow by the archer).

With this Gemara we perhaps can suggest an explanation for the significance of the flame miracle that took place in Kislev. The Gemara says that a flame is an arrow. Just as an arrow must have been shot by an archer, the flame of Chanukah could not have burned for eight days without having been lit by the omnipotent Creator. Kislev is the time when we focus on the Archer until we recognize that everything in this world comes from His Hand.

Refined Liquid and Refracted Light

In Kislev, the winter stretches on and the days grow ever shorter. As we know, each month contains special and unique lessons from Hashem to inform us on how to live. Kislev is no exception. Let's analyze the unique characteristics of the month and see if we can decipher the hidden message.

Let's start with the most obvious uniqueness of our month—the holiday of Chanukah. We are all undoubtedly familiar with the story of how the Syrian-Greeks campaigned to assimilate us into the secular culture and how the Maccabees valiantly fought to defend our Torah worldview. We are also familiar with how the Maccabees rededicated the desecrated Temple, and how the sole remaining flask of pure olive oil miraculously burned for eight days.

However, if we are to gain a true understanding of Hashem's message, we must ask the following two questions: First, why did Hashem communicate His loving Presence specifically with the flask of oil? Secondly, why did Hashem arrange for this to occur in Kislev? The first question is not too difficult. There are numerous sources (see *Vayikra Rabbah* 36:3) that explain how oil is the symbol of Torah. Since the Maccabees fought to defend the Torah, Hashem communicated His approval with an oil-related miracle. But our second question still stands. We know that events only occur in months that are appropriate for them. In what way is Kislev the month of Torah? How is oil related to Kislev?

To answer this question, we need to understand the *mazal* (constellation) of the month—*Keshes.* In English, *Keshes* is referred to as Sagittarius, the Archer. This is because *Keshes* literally means a bow—the weapon of choice of an archer. Let's ask ourselves: Where does the Torah discuss a *keshes*? The clearest example can be found in *Parshas Noach.* After Noach and his family leave the Ark, Hashem promises that He will never again flood the entire world. Hashem declares, *"I have set My* keshes *(rainbow) in the cloud, and it shall be a sign of the covenant"* (*Bereishis* 9:12). A rainbow is called a *keshes* because it is shaped like an archer's bow. But how can a weapon be a symbol of a covenant of peace? The Ramban on this *parshah* explains that a warrior who wishes to display his peaceful intent will unstring his bow and reverse it so it points in the opposite direction of his opponent. Similarly, Hashem indicated His goodwill with an unstrung bow which He pointed heavenward (away from the earth).

So now we know why Hashem chose something in the shape of a bow. But why did He choose something so bright, beautiful and multi-colored? What is the significance of this aspect of the *keshes*?

Let's turn to the *Midrash Rabbah* in *Parshas Noach* (35:3): *[Why did Hashem choose a rainbow? Because] it is similar to Him. Is it really possible to compare to Him? [Obviously not, but we can still say that Hashem has a rainbow-like quality, even though] there is nothing that can truly compare to Hashem any more than a husk can compare to the grain.* This midrash is enigmatic. What rainbow-like quality does Hashem have? The commentators explain that this is a reference to Hashem's many *middos* (qualities). Although Hashem is ultimately One, we as humans perceive Him as if He has many distinct qualities. He is merciful, kind, just, mighty, humble, generous, forgiving, etc. Just like a single beam of light can refract into many colors and form a rainbow, so too Hashem's Oneness manifests in many radiant ways. The rainbow is perfect for the covenant of peace because any time the people of the world deserve to be destroyed, all Hashem needs to do is "look back" at *His middos*—and we can rest assured that this will result in benevolence.

So where are we holding? We explained that *Keshes* is an archer's bow and why Hashem chose a weapon to represent His treaty of peace. We also discussed how the many colors of the rainbow correspond to the many *middos* of Hashem and how that makes the *keshes* an even better choice for the treaty. But we still don't know what this has to do with us. What message can *we* glean from the rainbow? Our journey is not yet over—let's take another two steps.

Step one is to the *Midrash Rabbah* in *Lech Lecha* (44:1): *[Does Hashem gain anything when we perform mitzvos? Of course not! He is already perfect. Rather,] mitzvos were only given so they would purify us.* This extremely fundamental midrash is teaching us that the mitzvos were only given so we could incorporate G-dliness into ourselves and become holier people. Each mitzvah incorporates a different G-dly quality into our souls. Some mitzvos make us kinder while others make us humbler. Some make us more generous while others make us more disciplined. The overall goal, though, is to integrate the *middos* of Hashem's rainbow until we shine with our own rainbow of good and holy qualities. That's step one.

Step two brings us to the *Derech Hashem* (part 2, ch. 3). The Ramchal writes that the reason Hashem gives different people different life settings is in order to grant each individual the ability to achieve his unique purpose: *If wealth or poverty did not exist, there would be no opportunity for people to demonstrate their generosity. The rich are tested by their advantage, while the poor are tested with regard to if they will be satisfied and thank G-d for the little they have.*

This fundamental quote is teaching us that each person has a different purpose in this world. While we all have the job to incorporate as many G-dly qualities as possible, each individual can practically only work on a limited number of them. A rich man can work on his generosity but not on being satisfied with little. Conversely, a poor man can grow in his satisfaction with less but simply does not have the means with which to be generous. The result is that, practically, no one individual can create the rainbow of G-dly qualities. Only together can we produce all the colors.

Now we can backtrack and answer our questions. Why is Kislev the month of both oil and Torah? I think the answer is because only with oil can we support a flame and create light. Hashem can make a sun that creates a rainbow, but we humans need oil to enable illumination. If we are to create a rainbow of holy and G-dly qualities, we must do it through Torah. There is no other way. While it is true that our differing abilities and life settings enjoin each of us to act differently, we all must work within the same Torah parameters. While we all live our lives on personalized paths, if we all follow the same guidelines, we can produce those individualized radiant beams that will join to create the rainbow that our nation is meant to project.

Chanukah: The Secret of Eternity

As the nights grow longer and darker, we can all use that rejuvenating injection of holiday spirit that comes with the arrival of Chanukah. Who doesn't enjoy recounting the story of Yehudah Hamaccabee and his valiant brothers who dared to challenge the mighty Greek Empire and win battle after battle against all odds? Which Jewish woman doesn't feel a sense of pride when relating the account of the virtuous Yehudis who infiltrated the enemy camp to assassinate their leader while he slept? Can one help but share the sorrow of the *kohanim* as we recall their entry into the Temple, only to find it utterly desecrated, save one lone jug of pure oil? Don't we all feel the loving Hand of G-d as we tell of how one day's worth of oil miraculously burned for eight days? What a wondrous tale! Such an agonizing struggle, but the triumph of victory is sweet. We can easily spend eight nights reveling in our glorious history.

But therein lies the trap. Yes, we can all too easily celebrate Chanukah in terms of the Chanukah story. But is that what Chanukah is all about? Did our Sages enact an eight-day holiday merely so that we could spend its duration reliving the past? Is Chanukah really only about events that happened long ago, and men and women long since gone? While I do not want to negate the value of studying our illustrious history, I believe the answer is an emphatic "no!" One who celebrates Chanukah (or any holiday) by merely commemorating past events is robbing the holiday of a tremendous amount of meaning and, quite frankly, selling himself short. Let us probe the story of Chanukah together and see if we can't discover that the war that was fought then is a war that is still being waged in our days. Let us see if we can uncover the secret of the Maccabees' victory and, thus equipped, ensure our victory as in those days (*bayamim haheim*), so too in this time (*bazman hazeh*).

Let's begin by dispelling a common misconception. The Maccabees were not fighting a war of independence. At no point in the time of the Second Commonwealth was Israel an independent state, before the Chanukah story or after. Throughout the era of the second Beis Hamikdash, the province of Israel changed hands numerous times. It went from the Persians to the Greeks and eventually to the Romans. The great Davidic Empire was long gone, destined to not rise again until the Messianic Era. The war of the Maccabees was not a war for independence. It was a war between philosophies, a war between two opposing religions. The Maccabees were not warriors—far from it! They were priests in the Temple of G-d, guardians of the Covenant. The war they were fighting was over the right to live a Jewish life the way we have always lived since we received the Torah at Sinai so many years ago. The Greeks were not trying to kill our bodies; they wanted to kill our souls. They wanted to destroy our holy way of life. The Greek philosophy was "Eat, drink, and be merry for tomorrow we die." Hedonism was the loftiest of ideals. They said, "Lusts are meant to be fulfilled! We're here for pleasure!" The body was idolized and it was believed that if one was strong enough, he could challenge the gods.

One can easily see how each of these ideals is diametrically opposed to Judaism. The Torah teaches how This World is but a corridor to the Next World. We're only here to prepare ourselves for the World to Come. We must become as G-dly as we can during our tenure on this planet, giving to others as G-d gives to us. Temporary pleasures are a distraction and should only be used to help in our greater goal of coming closer to the Creator. The body is but a tool to be utilized in the humble service of G-d. What a clash! Such mutually exclusive philosophies!

One does not have to be particularly perceptive to recognize that this battle is still being fought today. The world we live in maintains the Greek philosophy that this world is the main place. "Live for yourself—you only live once!" "Grab all those pleasures; they will make you happy!" "Don't ask yourself, 'What does G-d want from me?' but rather, 'What do *I* want from me?'" These messages are eerily similar to the Greek way of life.

So what is left for us, as Jews, to do? We must step up to the plate, as the Maccabees did, and shout out, "*Mi laShem eilai*—'Whoever is for G-d, to me!'" Will we bow to the culture of the world or will we raise up our banner and fight to live in the service of G-d like we've been doing for over 3,300 years? Granted, no one is trying to kill us. But it is harder to live for Judaism than it is to die for Judaism. How can we win this battle?

The solution lies in the Chanukah menorah. How is oil different than other liquids? The answer is that *oil doesn't mix*. When oil and water are put together, oil naturally rises to the top. We must take on the qualities of oil. We have our pure Torah, passed down in an unbroken chain. We must not mix it with foreign philosophies. No matter what the world presently believes, we must stick to our eternal truths and rise to the top.

Our job is to support the flame of Torah and shine as a "light unto the nations."

Perhaps that is the message of the miracle of the oil lasting beyond its natural lifespan. Every nation, empire, and philosophy eventually crumbles and is confined to the dustbin of history. But Hashem showed us long ago that if we take on the quality of oil and stay separate from the foreign philosophies of the world, our nation will merit living beyond the normal lifespan of a people. Only through being like oil will we merit being that eternal nation we are meant to be.

Principles, Priests, and Peer Pressure

Judaism is a religion of growth. The Torah teaches that one must constantly be growing throughout his lifetime. Every day is another chance to come closer to the Creator. Throughout the year, there are opportunities the calendar provides to strengthen one's religious self in different ways. Thus we find that the Jewish calendar as a whole is geared toward creating a strong, close relationship with the Almighty. Each Jewish holiday has a different message (or "theme," if you will) on which we focus for the duration of a proscribed number of days.

During Kislev, we celebrate the holiday of Chanukah. What do you suppose is the theme of Chanukah? We all know the story behind the holiday. The Hellenistic Greek Empire tried to do away with the Jewish religion; they wanted us to Hellenize and live like them. Matisyahu and his five sons formed a militia which carried out numerous guerrilla attacks on the mighty Greek army and won countless miraculous battles. After recapturing the Temple and finding it defiled, the Maccabees lit the Temple Menorah with the single remaining flask of pure olive oil—which miraculously burned for eight days until new oil could be prepared. The question is, what are we supposed to gain from this story? What type of growth is intended in the commemoration of this historical narrative?

Perhaps one could suggest that we should be inspired by the Maccabees' commitment to the Torah and likewise strengthen ourselves in Torah observance. That is possible; Torah observance can always use strengthening. But we already have a holiday to focus on Torah—Shavuos. Perhaps the theme of Chanukah is that G-d can (and does) perform miracles for His people. But for that we have Pesach. Maybe the message of Chanukah is that Hashem will never abandon His people and He will be with us even in the darkest of times. But that message is already encapsulated in Purim. So in which special area should we be growing on the holiday of Chanukah?

Let's suggest as follows. If the holiday of Chanukah had to be summed up in one word, I think the correct choice would be the word "principles." Why did the Maccabees fight? The Greeks did not want to kill them; they were perfectly content to live side by side with the Jews. What prompted the Maccabees to be so foolhardy as to dare to take on the army of the mightiest empire in the world? The answer is: principles. The Hellenist Greeks weren't after our bodies; they were after our souls. They wanted us to be a nation like all the nations. They were willing to let us live *with* them if only we would live *like* them. Unlike Pharaoh or Haman who fought us with the sword, the Greeks welcomed us with open arms. "Join us," they said. "Adopt our culture. Pursue our goals." What an opportunity it was! The nations were finally ready to accept us. All we had to do was accept their worldview.

Many Jews did just that. But a small group of *kohanim* held strong. Matisyahu and his sons understood that the Torah is the eternal Word of G-d. It is the Guidebook upon which we base our life. They stood up for the glory of the Torah and cried out "*Mi LaShem eilai!*—'Whoever is for G-d, to me!'"

The Maccabees were people of principle. From where did they draw the strength to face down the world power? The answer is, from an unswerving dedication to Hashem and His laws. It has been said that to live a meaningful life, one must find out what it is he is willing to die for and then go live for it. The Maccabees did just that. They raised up the banner of

Torah, and all those loyal to Judaism rallied to their flag. The rebellion was made up solely of people fighting not for their lives, but for their *way* of life. They fought for what they believed in. They fought for what was right. The miracle of the oil, occurring when the curtain had all but closed on the Age of Miracles, shone forth as an eternal message that courageous heroes who unstintingly stand with G-d and His ways will always merit His protection.

This message has a tremendous amount of relevance. Let's digress slightly to bring out this point. Around this time of year, we read the *parshah* of *Vayishlach*, which the Sages nicknamed *Parshas Hagalus* (the Chapter of Exile). This means that these verses describe the proper conduct for a Jew while in exile. On a deeper level, the nickname also alludes to the fact that all that will occur during *galus* is hinted to in this section. In the parlance of the Sages this is called *ma'aseh avos siman l'banim*—"the events [that occur to] the Patriarchs foreshadow [events that will occur to] the children [of Israel]."

In this *parshah*, after many years of separation, Yaakov Avinu confronts his brother Eisav. Their relationship was never too good, to say the least; it mainly consisted of Eisav constantly trying to kill Yaakov. In this confrontation, however, Yaakov finally manages to appease his wicked brother. No longer interested in fratricide, Eisav makes an astounding offer: *"Let us go and travel [together]; I will go alongside you"* (*Bereishis* 33:12). After many years of hatred and ill will, Eisav opens his arms wide and proposes to live side by side with his brother. And what is Yaakov's response? He makes excuses and tries to wiggle out of it! Yaakov says that his family can't travel fast enough to keep up, the sheep move too slowly, and *"I'll [eventually] catch up with you [another time]."* Yaakov had the wisdom to know that two opposing worldviews cannot live together. It is impossible to remain uninfluenced by one's surroundings. The creation of the Chosen Nation is a weighty responsibility and does not allow for any mixing with foreign philosophies.

Is it not obvious how this Torah portion foreshadows the times we are living in? We are an old people with a long history.

Through much of it, the nations were at our throats. Any method of murder that was invented was experimented with on our people. *Baruch Hashem*, we now have more religious freedom than we've ever had in all our history. The nations let us live side by side with them. The world at large is much more tolerant of us in the present day and age. Yet assimilation is at an all-time high. A very large percentage of Jews have accepted Eisav's invitation to live side by side with him. They cast off the yoke of Heaven and not only live *with* the nations, but also *like* the nations.

Perhaps this is an area in which we can strengthen ourselves on Chanukah. Are we people of principle or not? We have values and a value system—the Torah. Are we so lacking in moral fiber that we let down our guard the moment Eisav lowers his sword? We can do better than that. We are the Chosen People and we have both the responsibility and the ability to choose. Let's use Chanukah as the opportunity to grow in our ability to stick to eternal G-d-given principles. As the Torah commands, *Choose life!* (*Devarim* 30:19).

Now let's move on to a different part of this discussion. What does it mean to be people of principle? What is the defining feature of a strong person who adheres to a value system? The answer is the ability to stand up to peer pressure. Peer pressure is possibly the strongest social power, and it affects every person. Statistics show that the average person believes what everyone else believes, and morals are often decided by the masses. How many of us would have the courage to stand up to the crowd and correct an injustice being perpetrated? Are we strong enough in our beliefs that we are willing to defend them against onslaught? There are not too many people who are.

Avraham Avinu was one such individual. Having recognized the Creator, he set out to educate the entire world who believed otherwise. He debated with kings and theologians and refused to back down from what he knew was true. Can we imagine the strength of character that Avraham had and his sheer determination to stick to his principles? What a hero!

It's not easy to be this way. We often hear people saying, "But everyone is doing it." It's so much easier to follow the crowd. The Torah describes how Yaakov worked seven years for Lavan to be able to marry Rochel. However, Lavan hid his older daughter Leah under the veil instead! When confronted by Yaakov, Lavan replied, *"Such is not done in this place—to give the younger before the older"* (*Bereishis* 29:26). What a pathetic excuse! "But everyone is doing it." So what? What happened to the explicit agreement to give Rochel? Hiding behind the masses is cowardly.

Let's contrast that to when Dinah was abducted by the local Canaanite prince and violated. Shimon and Levi stood up and wiped out the city for, as they said, *such a thing is not done!* (*Bereishis* 34:7). Shimon and Levi didn't bow to societal standards; they *set* the standards. This is the mark of great men.

The heroes of the Chanukah story, the Maccabees, excelled in this trait as well. Their rallying cry was "*Mi laShem Eilai!*" From where did they learn this phrase? This expression was actually hundreds of years old, first used by their great-great-uncle Moshe Rabbeinu. When Moshe came down from Har Sinai and saw the people worshipping the Golden Calf, he called out, "*Mi laShem eilai!*" The Torah records: *And he (Moshe) called out, "Whoever is for G-d—to me!" and all the children of Levi gathered unto him* (*Shemos* 32:26). The Maccabees of the Chanukah story were *kohanim*, descended from the tribe of Levi. They learned this holy refrain from their ancestor Moshe and once again stood up to sanctify the Name of G-d. Chanukah is not about remembering these righteous men; Chanukah is about *emulating* these righteous men. If we can learn to stand up for eternal truths and principles, the holiday will be a success.

There is a famous story that brings out this message beautifully. Rabbi Shimon Schwab once went to visit the saintly Chafetz Chaim. At one point, the Chafetz Chaim asked him, "Are you a *kohen*?" Rabbi Schwab answered in the negative. The Chafetz Chaim responded, "I am a *kohen*, maybe you have heard?" "Yes, I have heard," responded Rav Schwab.

"Are you a *levi*?" the Chafetz Chaim then asked. Again, Rav Schwab answered in the negative. "What a pity!" said the

Chafetz Chaim. "Mashiach is coming and the Beis Hamikdash will be rebuilt. If you are not a *kohen*, you can't go in to do the *avodah*. Do you know why? Because three thousand years ago, by the incident of the Golden Calf, *dein zeide iz nisht geloffen*, your grandfather didn't run when Moshe Rabbeinu called out '*Mi laShem eilai*.'"

The Chafetz Chaim then concluded, "Now, take heart. When you hear the call *mi laShem eilai*, come running!" (*Selected Writings from Rav Schwab*).

Let's conclude by using this story to shed light on an enigmatic *pasuk*. Bnei Yisrael were gathered around Har Sinai, primed for a revelation unmatched to this day. Hashem came and chose us as His eternal nation. The verse reads: *And you shall be for Me a kingdom of priests and a holy nation* (*Shemos* 19:6). The part about being a holy nation is self-explanatory. But what does it mean to be a "kingdom of priests"? Only a portion of the Jewish people serves in the Temple. In what way does this charge apply to the rest of the nation? By now, the answer should be obvious. We are not only the physical descendants of Avraham Avinu; we are also his spiritual inheritors. As individuals, and collectively as a nation, we have the ability to stand up for the truth and hold strong no matter what. We have the ability to tap into that priestly strength demonstrated by the tribe of Levi in the time of Moshe and by the Maccabees in the time of Matisyahu.

Let's use Chanukah as the time to revitalize our pride in being the Chosen People and remind ourselves that we bear this title because we have the ability to choose. We are a kingdom of priests. Let's not submit to the peer pressure of Eisav; the light unto the nations cannot shine when mingled with foreign values. Let's join together and raise the banner of Torah. *Mi laShem eilai!*

Cows, Greeks, and Little Flames

What do beautiful cows, the Greeks, and small flames have in common? Seemingly, not much. Yet every year without exception, we kindle small flames to celebrate the miracles we experienced in the days of the Greek Empire specifically during the week of *Parshas Mikeitz*, when Pharaoh had his dream about the seven skinny cows and seven fat cows. "So what?" you may ask. "What connection can you possibly draw between these unrelated topics?" Come join me as we set off to discover that the fact that Chanukah coincides with this *parshah* is not a coincidence after all.[2]

Let's begin our analysis with *Parshas Mikeitz*. The verse describes how Pharaoh first saw the representations of the seven years of plenty (the seven fat cows) and then saw the representations of the seven years of hunger (the seven skinny cows). But let's analyze the precise wording used to describe the fat cows (*Bereishis* 41:2): *sheva paros yefos mareh*—"seven cows that were beautiful of appearance." Why does the verse use the term *yefos mareh*—"beautiful of appearance"? Why not just say they were *yafos*—"beautiful"? To address this question, Rashi steps in and explains: *[This lengthier term is the more precise way to describe] the days of plenty because [during days of plenty] people appear beautiful to one another.*

What does Rashi mean when he says that people can at certain times "appear beautiful"? Ostensibly, one is either beautiful or not beautiful! How can an objective fact such as beauty change depending on the economic situation of the world at large? Presumably, the explanation is as follows: There is no such thing as someone or something being objectively beautiful. Every creation has positive characteristics and negative characteristics. Everyone has ways in which he or she is beautiful as well as ways in which he or she is ugly. If you choose to look at my positive qualities, I will appear to be beautiful; if you choose to look at my negative qualities, I will appear to be ugly. As such,

---

2. I don't claim credit for the idea that Chanukah relates to the *parshiyos* that we read at that time. The *Pri Tzaddik* (*Ma'amar L'Chodesh Teves, os* 17) quotes the Shelah Hakadosh that: *The* parshiyos *of* Vayeishev, Mikeitz, *and* Vayigash*... fall out during Chanukah, or a bit before or after. Certainly, then, they are connected...* See there for his Kabbalistic explanation of this phenomenon.

in times of plenty—when people are happier and more predisposed to focus on the positive aspects of their fellows—people will appear to be beautiful. Conversely, when times are tough and people are more predisposed to focus on others' negative traits, the people of the world will appear ugly to one another. In short, beauty is not objective, but rather contingent on the eye of the beholder.

Now let's connect this idea to the Greeks. This Hellenistic Empire was known for its glorification of all kinds of beauty. They venerated beauty in architecture, music, drama, sculpture, and, above all, in the human body. Now, with the advent of any new idea or philosophy, every Torah-observant Jew has the responsibility to ask himself, "What is the Torah's view regarding these ideas?" We have long remained true to Hashem's path only by constantly checking how the Torah wants us to respond to all that the world throws our way. So when the Greeks introduce the ideas of an aesthetic-based value system to the world, our first step should naturally be to open up the Torah and see what we can glean.

Voila, at the very beginning of *Parshas Mikeitz* we have our discussion about beauty! And what does it say? It says that there is no such thing as objective beauty. Anything and anyone can be beautiful in your eyes if you choose to focus on the positive qualities. And anything and anyone can appear horribly ugly if that is how you choose to view them. Our suggestion herein is that *Parshas Mikeitz* falls out during Chanukah so that when we wonder about the Torah's approach to beauty, we can immediately find our answer in the weekly portion.

Let's conclude with one final thought: There is a law codified in the *Shulchan Aruch* (O.C. 671:3) that states: *If one fills a bowl with oil (on Chanukah) and places many long wicks in it and lights them...he has not fulfilled his obligation [to light many flames, nor does he even fulfill his minimum obligation] to light one single flame.* This is an interesting halachah. It is not too difficult to understand why one doesn't fulfill the obligation to light many flames in this way. After all (as the *Mishnah Berurah* explains), the flames join together into one united flame and therefore cannot be considered to be many distinct flames. But if there is at least one (large) flame, why isn't the minimum obligation (to light one flame) met? Apparently, a big flame is not a fulfillment of the mitzvah. But why is this? Why can one only do the mitzvah with a small flame?

Perhaps we can suggest the following homiletical thought: We learned in the *parshah* that comes out on Chanukah that beauty is not objective, as the Greeks believed; it is rather in the subjective control of each individual. Each person can choose the qualities on which he wishes to focus, making beauty something that is less of an absolute. Similarly, we can explain that when one has a large flame, everything is noticeable. The large light makes all details visible. However, a small flame can only be focused in one direction, and one must necessarily choose those qualities on which he wants to focus.

Let's learn this lesson on Chanukah—to view beauty like a Jew and not a Greek. The Greeks taught that beauty is an absolute and can therefore be venerated. The Torah's account of the cows teaches us that everyone has qualities to admire, depending on where we place our focus. We retain the power to make the world a more beautiful place, if we only choose to do so.

Teves

The Dancing Goat

As the last of the Chanukah flames flicker out, the month of Teves comes upon us. What special qualities does the month of Teves have? In which area can we grow? Let's begin an analysis of Teves and hopefully learn something that will help us become bigger and better people.[1]

Let's commence by turning to the faculty of the month. The *Sefer Yetzirah* writes that this is *rogez*—we will not translate this word because it is completely impossible to find an English rendering that would connote all of its possible meanings.

Let's quote a few *pesukim* to bring out this point: 1. The prophet Chavakuk (3:2) pleads with Hashem, "*B'**rogez** racheim tizkor.*" This translates as "Remember to be merciful even during [Your] **wrath**." 2. In *Tehillim* (4:5), King Dovid advises: ***Rigzu** v'al techeta'u*, which roughly means, "**Tremble** [before G-d's Ever-Presence], and thereby avoid sin." 3. We find Yosef exhorting his brothers (*Bereishis* 45:24), "*Al tir**gizu** baderech*—'[Go straight back to tell Yaakov that I am still alive.] Do not become **agitated** on the way.'" Rashi explains that this means, "Do not become caught up in discussing a Torah matter [because it will cause you to be delayed.]"

Thus, we see that the word *rogez* can connote wrath, trembling, and even fierce discussion over a matter of Torah. This is very

---

1. The following essay is almost completely based on *Pri Tzaddik* (*Ma'amar L'Chodesh Teves, os* 17). There he writes about A) how Teves is the month of the *kaveid*, liver, which the *Zohar* says relates to Eisav because it's filled with blood, and there must be a good way to use this part of us, B) how *rogez* is a dual *middah* (character trait) that can be used for anger or fear of Heaven, C) how there are different *pesukim* that bring out this duality of *rogez*, and D) how the Raavad explains that *gedi* is a *se'ir* and a *se'ir* is Eisav and a demon, etc.

confusing. How can all these things be part of one faculty? What do they have in common? And what can we learn from it?

Perhaps we can find a hint by taking a detour to another characteristic of the month. Each month is aligned with one of the *mazalos* (constellations), with Teves matching up with *Gedi*—also known as Capricorn, the Goat. What do we know about goats? Well, one interesting point is that goats are called by a few different names in the Torah—*gedi, se'ir,* and *eiz*. While we will not be discussing the differences between the names, it may be worthwhile to note that the Hebrew word *se'ir* not only means a goat but also a demon. What is the connection between goats and demons? Let's take a look at the prophecy in *Yeshayah* (ch. 13) regarding the destruction and subsequent desolation of Babylon: *Babylonia, the choicest of the kingdoms, will...be like Sedom and Amorah.... It will not be inhabited... Their houses will be full of ferrets...and* se'irim *will dance there.*

What are these *se'irim* that will dance amidst the ruins? The Radak explains: *These* se'irim *are demons. [But why did the* pasuk *choose the word* se'irim *instead of the more common word for demons,* sheidim*? The answer is because] the demons will appear in the form in which demons commonly appear—the form of goats.*

Essentially, Yeshayahu is prophesying that the ruins of Babylonia will be overrun by demons that will appear in the form of dancing goats. This discussion leaves us with two potent questions: Why would goats dance? And how does this (apparently) goat-like quality relate to Teves?

To answer this question, let's embark on a short analysis of dancing. Why would someone dance? Is it not an illogical course of action? What goal is accomplished by agitating one's limbs to follow a rhythm? The answer is that a person does not dance to accomplish a goal, but rather because his spirit moves him to do so. When one's spirit is stirred up in a joyous way, it manifests that arousal in what we call a dance.

This discovery is quite important for two reasons: First, it informs us that goats dance obviously because they have an active spirit coursing through them that rouses them to skip and leap as they do. Second, and far more important, is that since humans also dance it must be that humans also have that same wakeful spirit within them.

We began with the perplexing question regarding what *rogez* means. How can it simultaneously refer to wrath, trembling, and getting caught up in a fierce Torah discussion? Perhaps now we can explain that *rogez* refers to the arousal of the spirit latent within each and every one of us. We know that Hashem invested us with more than just physical qualities. Although we cannot see or measure it, one of the non-corporeal aspects of the human is his spirit that, according to our Sages, resides in the blood. When this spirit is aroused, it can manifest in rage and anger. Alternatively, if one focuses on how Hashem is in front of Him at every moment, this spirit reacts with fear of G-d and trembling. When one engages in a fierce, competitive debate over a matter of halachah, this spirit gets excited over the battle. But no matter in which way the spirit is moved, this arousal of the spirit is called *rogez*. Incidentally, this explanation of *rogez* also helps us understand *Midrash Rabbah* (84:19) that says that a goat's blood is similar to a human's blood. After all, don't both bloodstreams carry the same active spirit?

It seems we have answered most of our questions. We elaborated on the meaning of a dance and discussed why one would say that goats are dancing creatures. We also managed to explain the meaning of *rogez*. But we are not yet done. What does this all have to do with us? If the *Gedi* is aligned with Teves then it is logical to assume that the goat's unique quality is the one we should improve during this time. So how can this be done?

The answer is that we are supposed to be aware of this spirit within us so we can use it properly. Eisav was a wicked man who was quite in touch with his spirit. Indeed, Eisav was called *Se'ir.* But Eisav would manifest his spirit in rage and anger. Yosef's brothers, on the other hand, would discuss halachah and tap into that animating force in their blood to bring forth the fierceness that can contribute so much to a Torah debate. In *Tehillim*, Dovid Hamelech exhorts us to arouse our spirits to tremble before the King of kings. There are so many good ways to use our spirit, but it can only be done if we recognize that we have it and work to use it properly.

May we be blessed to indeed connect to this force with which Hashem endowed us and to succeed in channeling it in all the right ways.

Snake Eyes

One of the many unique characteristics of the month of Teves is the letter with which it is aligned. According to the *Sefer Yetzirah*, that letter is *ayin*. The commentators elaborate on this by saying that Teves is the month of the eye since the Hebrew word for eye is *ayin*. Additionally, *Pri Tzaddik* (*Ma'amar L'Chodesh Teves, os* 17) quotes the Kabbalah that Teves is the month of *eynin d'Malka*—which means something along the lines of that this month is the month of eyes. Hence, our job this month is to discover what we should fix regarding our eyes.[2] In addition, we need to discover why Asarah B'Teves falls out in this month. What is our *avodah*, specific job, for this time of year? Come; there's a long road ahead.

Let's take our first step by turning to the tribe of Teves, which is Dan. When we want to learn about the tribal qualities, we look to the blessings of Yaakov Avinu and Moshe Rabbeinu. Yaakov says, *"Dan will avenge his people... Dan will be a serpent on the highway..."* (*Bereishis* 49:16). The commentators explain that this prophecy is a reference to Shimshon—the righteous leader who fought for the Jewish people during the days of the Judges. So we first need to ask, what is the connection between Dan and the snake? Second, in what way is it appropriate for Shimshon

---

2. *Pri Tzaddik* (*Ma'amar L'Chodesh Teves, os* 17) says: *The main thing in [Teves] is to fix ourselves regarding the matter of* shemiras einayim *(guarding one's eyes), like the* pasuk *says,* v'lo sasuru acharei levavchem v'acharei eineichem...—*"and you shall not turn after your heart and after your eyes."* The following essay is an exposition of that short comment.

to have descended from the anguine tribe? Third, how do Shimshon and the snake connect to Teves and Asarah B'Teves?

Let's take the next step by analyzing the story of Gan Eden as recorded in *Sefer Bereishis*. We all know that the serpent enticed Chavah to eat from the tree. But *why* did the serpent do so? What did he hope to accomplish? The answer can be found in Rashi (3:1): *The Torah juxtaposes the story of the snake with the* pasuk *that states that Adam and Chavah were unclothed to explain that [the snake] chose to act because he saw [Adam and] Chavah unclothed and…he desired her.* Rashi is explaining that the snake wanted Chavah to feed the forbidden fruit to Adam so he would die, thereby leaving Chavah available for the snake's advancements.

If we read the Rashi carefully, we'll see that the snake only desired Chavah because he saw her unclothed. The *desiring* only came as a result of the *looking*. We see the same concept in the Shema: *You shall not stray after your heart and after your eyes.* Rabbeinu Bacheye explains: *The* pasuk *only says "heart" before "eyes" because the heart is the more prominent organ. However, in reality, when one sins it is because first the eye sees and only then does the heart desire.*

What this means for our discussion is that the snake didn't have an intrinsic desire for Chavah; it only began after he looked at her. If he would have guarded his eyes and hadn't looked at what wasn't befitting for him, he wouldn't have had an issue. This was Shimshon's undoing as well. The mishnah (*Sotah* 9b) states: *Because Shimshon strayed after his eyes, [Hashem allowed] the Philistines to gouge them out.* Although Shimshon was a great tzaddik (righteous person), he strayed after his eyes to a degree inappropriate for a man of his stature. Because he gazed at that which was not fitting for him, his judgment became slightly clouded and he strayed after foreign women. This caused his ultimate downfall.

This is quite a powerful concept. We are only human, with base desires and selfish wants. But it is possible to avoid these lowly cravings altogether—by not looking in the first place. There is no good that can come of looking at what your friend has. A story comes to mind about a great Yerushalmi *rav* who related that he remembered only one time that his mother ever smacked him. His mother was cutting pieces of cake for him and his siblings and he looked over to see how big his brother's piece was. His mother smacked him and scolded, "In life, you don't look at somebody else's plate."

So we've answered the first two questions. We understand that the tribe of Dan is compared to a snake because they suffer from the negative snake-like tendency of gazing at what is not befitting, and we understand that even the great Shimshon succumbed to his tribal predilection. All we have left is the third question—how does this connect to Asarah B'Teves?

I would like to suggest the following answer: Perhaps the beginning of the siege (on Asarah B'Teves) which led to the destruction of the first Beis Hamikdash was only made possible because the Jewish people strayed after their eyes. Now, I know the Gemara that says that the first Beis Hamikdash was destroyed because of the three cardinal sins (*Yuma* 9b). But how did the holy people of Israel come to adopt the degenerate ways of the nations? What could possibly have caused this to happen? It was only because they strayed after their eyes.

Let's prove this by taking a detour to *Parshas Nitzavim*. The Jewish people stand primed to enter Eretz Yisrael, and Moshe gathers them to renew their covenant with Hashem. Moshe explains, *"You are standing here today...before Hashem...to pass into the covenant. [Why is this covenant necessary?] Because you know how we dwelled in Egypt... Additionally, [during our travels, you] passed through the midst of the nations. You saw their abominations and idols. [So] perhaps there is among you...one whose heart turns away from Hashem...to go serve the gods of the nations..."*

Is what Moshe is saying not incredible? It seems that the same people who saw the plagues, the splitting of the Red Sea, and the revelation at Sinai still might be seduced to follow the ways of the nations. Why? Because they *saw* the ways of the nations. The evil inclination can convince you to do all sorts of things, but only once you *look* and let the abomination into your life. In this fashion, I would explain that the Beis Hamikdash was destroyed because the Jewish people strayed after their eyes and thereby enabled the ways of the nations to seep into their culture.

However, everything can be used for both bad and good. It is wrong to look at our friend's house, car, and spouse, but it is commendable to look at the world and see the wonders of creation. Hashem made so many incredible things; if we would only open our eyes, we would be constantly astounded and awed.[3] I would even suggest that perhaps a reason Hashem made so many wondrous creations is so that we can accomplish *shivisi Hashem l'negdi samid*—seeing Hashem wherever we go. If we work to see Hashem's wonders, then we will see Hashem everywhere we look—and we won't dare look where we shouldn't.

Let's conclude with suggesting one final thought: The midrash in *Vayechi* teaches that Shimshon could have been the Mashiach. This being the case, I think we can suggest that the fact that he failed because he strayed after his eyes is not random. The Beis Hamikdash is called the "Eye of the Universe" (*Bava Basra* 4a). Someone who doesn't use his eyes properly cannot build the Eye of the Universe. The Jewish people collectively strayed after their eyes and consequently the Eye of the Universe was taken from them. If we can use Teves to use our eyes properly and work on not looking at what we shouldn't, we'll merit seeing Mashiach, soon and in our days.

---

3. Indeed, the quote from the *Pri Tzaddik* in the previous footnote concludes: *Rather than the eyes being used for straying, we should us them for* se'u marom eineichem—*to lift our eyes to see the wonders of Hashem.*

Shevat

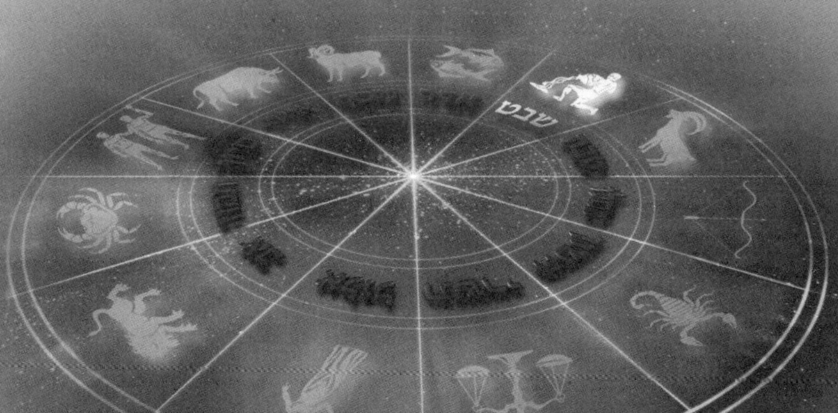

Fruits, Buckets, and Holiness

Judaism is a religion of growth. Hashem wants each individual to perfect himself in as many ways as possible. However, it is impossible to focus on everything at one time. Therefore, the Torah provides us with a calendar, which compartmentalizes different areas of growth into separate months. Each month, with a holiday as its focal point, is ordained as the time for us to work on a particular aspect of our service to Hashem.

The month of Shevat is no exception. But what is the growth of Shevat? Tu B'Shevat, the New Year for the Tree, is scarcely even a holiday; the only observance of the day is found in the comment of the *Magen Avraham* (O.C. 131:16): *It is the custom to eat many types of fruits on Tu B'Shevat.* What type of holiness can be found in eating fruits? Is there anything more mundane than eating? Even if the requisite *brachah* is recited, what is so special about eating fruit with a blessing on Tu B'Shevat (or during Shevat)?[1]

Before attempting to answer this question, let's add another question to the mix: In Jewish thought, each month has a *mazal* (constellation) that embodies the essence of the month. The *mazal* of Shevat is *Dli*, which literally translates as "the Bucket" but is more commonly known as Aquarius, the Water Bearer. Interestingly, not only is *Dli* the *mazal* of Shevat, it is also the *mazal* of Am Yisrael—the Jewish people.[2] How can we understand these enigmatic ideas? What does Shevat have to do with buckets? Why is the kingly and precious nation of Klal

---

1. The *Pri Tzaddik* (*Ma'amar L'Chodesh Shevat, os* 1) says: *The job of this month is* l'hachnis kedushah b'achilah–*to bring holiness into eating.* This essay will try to explain how precisely that can be done.

2. I found this idea in the *Bnei Yissaschar* (*Ma'amar* 1 *L'Chodesh Shevat*), where it is quoted from earlier sources.

Yisrael compared to a bucket whose value is only defined through its function as a tool to contain something of greater value?

Let's see if we can take a detour to gain the understanding needed to answer these questions. Why did Hashem create us with a need to eat? Wouldn't life have been so much simpler if we didn't need food? So much time and money could be saved, and we would be that much freer to serve Hashem. The answer can be found in the *Chovos Halevavos* (*Sha'ar Habitachon*, ch. 3): *Hashem's wisdom decreed the testing of the soul if it will serve Hashem or rebel against Him. The manner in which it is tested is in regard to the soul's deficiency and need for externals such as food and drink. Hashem commanded the soul to seek out the externals in order to see how they will be sought...* What this means is that eating is not an unfortunate prerequisite to serving Hashem; eating is actually part of serving Him!

However, this is something that requires explanation. How can the self-serving act of eating be considered a mitzvah? I think the answer lies in understanding what exactly a mitzvah is. The Ramban writes at the end of *Parshas Bo*: *All mitzvos are extremely precious, for through them a person comes to acknowledge his Creator. Indeed, the objective of all mitzvos is that we will come to acknowledge our Creator, that He created us.*

It seems from the Ramban that everything in the world can be used for a mitzvah if it is used to turn our hearts and minds toward Hashem. This calls to mind a story of a chassid who was traveling with his rebbe. By chance, they both took out an apple at the same time. The chassid jokingly commented that he was finally able to equal his leader. The rebbe gently rebuked him, "You think we are the same? You make a *brachah* because you want to eat. I eat because I want to make a *brachah*." Food can be eaten to fill your stomach. But food can also be eaten with the intent to recognize the creation and acknowledge the Creator. A *brachah* is not just a "thank you." It's an acknowledgment of the source of one's food. The *brachah* is supposed to remind a person to see Hashem in every bite. Look what Hashem created. The apple is so delicious. The orange is delectable. Where did it come from? Who made it? Hashem wants us to enjoy our food; that's why He made it taste good in the first place. But He wants us to use these delicious creations as a mitzvah—to acknowledge Him as the *Melech Ha'olam*, the Master of the world.

Perhaps now we can understand why we eat many different types of fruit on Tu B'Shevat. If the only growth in eating fruit is in the making of the *brachah*, it would suffice to eat only one fruit. Why would there be a need to sample a variety? The answer is clear. Exploring a great variety of Hashem's creations multiplies our recognition of His handiwork exponentially. Hashem not only created the apple and the orange, He also made the grapefruit and kiwi. He made grapes, olives, pomegranates, peaches, walnuts, hazelnuts, coconuts, avocados, papayas, almonds, dates, figs, apricots, lemons, mangoes, pears, cherries, and a few hundred other species of fruit. No one fruit has the same taste as the other! They are all so different, it would almost be logical to make a different *brachah* on each one, just as we don't make the same *brachah* on vegetables as we do on fruits.

However, we know it is the same *Melech Ha'olam* who created each one of these completely different fruits. The *Chovos Halevavos* (*Sha'ar Habechinah*, ch. 1) says: *If something always does one thing, that indicates that it is forced by its nature to do only that. Fire can only burn; water can only cool. But Hashem is omnipotent and is not forced by any nature. Therefore, one can see great variety in the creation*… Tu B'Shevat is the day for recognizing that variety. By sampling a great variety of creations, we naturally come to see their Creator.

Thus armed, I believe we can understand why *Dli* is the *mazal* of Shevat. As mentioned, a bucket has no intrinsic value. Its only value is as a tool. Similarly, fruits serve no inherent function. Hashem could have made us without the need for food. However, fruits are very valuable and precious because they can be used as a tool to focus us on the Creator. Fruits are the vessels (or buckets, if you will) made to contain holiness. When we make a *brachah* and eat each one, we acknowledge Hashem as the Master of the world.

This is also why *Dli* is the *mazal* of the Jewish people. Judaism is not about rituals. Nor is it about the synagogue. Judaism is about seeing Hashem in every aspect of our lives and acknowledging Him there. We take acts as mundane as eating and transform them into tools, into buckets overflowing with holiness.[3] *This* is the growth of Shevat.

Instant Life (Just Add Oil)

We all know that the uniqueness of the month can be alluded to by the special days it contains. However, the only special day in Shevat is Tu B'Shevat. What is the significance of this obscure day in the calendar? The first mishnah in *Maseches Rosh Hashanah* explains that Tu B'Shevat is the "New Year for the Tree." Rashi explains that the mishnah is teaching that the fruits that blossom before Shevat must be part of a separate tithe calculation from the fruits that blossom after Shevat. However, if this idea was all the Sages had in mind when they taught this mishnah, why did they say "tree" in the singular? Shouldn't they have said "New Year for the Tree**s**"? This grammatical anomaly implies that Tu B'Shevat also carries significance with regard to a specific tree. But which tree can possibly carry such significance to be awarded its own day? And why should that special day be in the month of Shevat?

When thinking about which tree is unique in the Torah's view, what jumps to mind is the most famous (or infamous) tree in history—the *Eitz Hada'as*. The Torah describes how Adam and Chavah were warned not to eat from the Tree of Knowledge. As we all know, they failed the test and were consequently expelled from the idyllic existence of Eden. Death was decreed upon the world's inhabitants. Adam would now eat by the "sweat of his brow" and Chavah would "bear children in pain."

3. The *Bnei Yissaschar* (*Ma'amar* 1 *L'Chodesh Shevat*) says this idea: Dli *is the* mazal *of Yisrael because a* dli *is made to draw water…it is only meant to serve for water…and water is Torah (i.e. spirituality).*

We all learned this story in kindergarten, but have we ever really understood it? Why were there such severe repercussions to their sin? Why was eating the fruit such an egregious crime? If we can understand this story, perhaps we can understand the significance of the Month of the Tree.

Let's take a detour. Each month in the Jewish calendar is aligned with a tribe, with Shevat being aligned with Asher. This is definitely fitting, as the symbol on Asher's flag was a tree. But what else do we know about Asher? Yaakov stated, *"Asher's bread will have richness, and he will provide kingly delicacies"* (*Bereishis* 49:20). It seems that Asher is the wealthy tribe who never suffers from lack. Moshe similarly said, *"The most blessed of children is Asher…he shall dip his feet in oil"* (*Devarim* 33:24). Oil is the symbol of prosperity.

The problem is that the Torah doesn't really take such a kindly view of being overly involved with the temporal world. The Ramchal writes in the first chapter of *Mesillas Yesharim*: *If a person allows himself to be drawn after the physical world, he will be distanced from the Creator.* The *Chovos Halevavos* writes similarly in the second chapter of *Sha'ar Avodas Elokim*: *The inclination toward the physical is always being flexed (like a muscle), since a person is constantly eating, drinking, and taking care of his body. Therefore, the inclination toward the physical is far stronger than the inclination toward the spiritual.* It seems from these sources that it is extremely important to *limit* our involvement with materialism. So where do Asher and his kingly delicacies fit into Judaism?

In order to answer these questions, let's tell a story: Yankel is a religious man who lives his life according to the Torah. At the ripe old age of 120, he dies and ascends to the Heavenly Court. "Yankel Friedman, step forward! Please confirm your identity. Are you Yankel Friedman?" "Yes." "You are Yankel Friedman of Brooklyn, New York?" "Yes." "The lawyer?" "Yes." "Yankel Friedman, the court is here today to reward you for the thirty-five years you spent in the service of the King of kings." Yankel interrupts, "There must be some mistake! I lived 120 years, not 35. You have the wrong file! I want reward for 120 years of service!" The angels of the court look over the file, but the information seems to be in order. It says that Yankel only spent 35 out of 120 years serving Hashem. "How can this be?" asks Yankel. The judge directs Yankel's attention to the video screen where many numbers pop up. "Yankel, you spent six hours a day asleep. That adds up to thirty years. You spent fifteen minutes on breakfast, a half hour on lunch, and forty-five minutes on supper. An hour and a half every day for 120 years adds up to seven and a half years. Subtract another seven and a half years for the hour and a half you commuted every day. Then take off forty more years for the eight daily hours of your nine-to-five job. That adds up to eighty-five years. All you have left, Yankel, is thirty-five years of service of G-d." Yankel sputters in disbelief, "What do you want from my life?! If I hadn't worked, I wouldn't have had food. If I hadn't eaten, I would have starved. And I had to sleep every day after a long day of work! What do you expect from me? You can't chop off eighty-five years of my life just like that!" The angel gently responds, "Yankel, at the end of the day, for eighty-five years you weren't serving Hashem; you were serving yourself."

Perhaps now we can answer our questions. If you involve yourself in physicality for your own sake, then you will indeed be distanced from your Creator. But it is possible to do all the above activities for Hashem's sake. Imagine stopping for ten seconds before eating a meal and thinking, "Hashem wants me to take care of my body, so I'd better eat." Imagine taking ten seconds before going to sleep and thinking, "I'm going to sleep so I can do what Hashem asks of me tomorrow." Imagine taking ten seconds before going to work and thinking, "I need to work so I can live my life the way Hashem wants me to." Boom! In thirty seconds each day, you just added eighty-five years of *avodas Hashem* to your life.[4]

Now let's backtrack to Adam and Chavah. The essence of

4. This lesson of Shevat is from the *Pri Tzaddik* (*Ma'amar L'Chodesh Shevat, os* 1): *The job of this month is* l'hachnis kedushah b'achilah—*to bring holiness into eating... This [can be done in the] way of the Rambam (*Hilchos De'ios *3, 3) who says one should eat in order to have strength to serve Hashem and sleep so that his*

their crime was that in a world that was created for service of Hashem, Adam and Chavah chose to use it to serve themselves. This is what the midrash (*Bereishis Rabbah* 16:7) means when it says that when Adam searched for a tree to give him leaves for clothing, all the trees shrank back and exclaimed, "Let not... arrogance come toward me!" It is the height of arrogance for a man who was given everything to selfishly serve himself instead of his Creator.

Let's tie it all together. Asher's kingly delicacies *do* have a place in Judaism. With proper *kavanah* (intent) a person can use all that this world provides in the service of Hashem.[5] This was all that Hashem ever asked of mankind—since the primeval command of the Tree of Knowledge. The *avodah* of Shevat is to learn the lesson of *the* Tree—to use the physical to serve Hashem. Additionally, perhaps we can add that this is why Asher was blessed with oil. While oil does represent prosperity and excess, it is also the only liquid that can properly uphold a flame. The only way to have Torah and mitzvos is through the physical world—and Torah and mitzvos are likened to a flame and a light [*Ner mitzvah v'Torah ohr*—"a mitzvah is a flame and Torah is light" (*Mishlei* 6:23)]. If we use Shevat as a time to learn how to use the physical world for Hashem's sake, we'll merit having used every moment of our lives in the service of the Creator.

The Mission of Submission

Hashem, in His kindness, gave each month its own special qualities so we can compartmentalize our growth into different

---

mind can rest; in this way his eating and sleeping will be considered actual service of Hashem.

5. *Pri Tzaddik* (*Ma'amar L'Chodesh Shevat, os* 1) says: *This month is aligned with Asher...because one can fix all and bring holiness into eating—which is the characteristic of Asher...*

parts of the year. Let's jump into Shevat and see what we can pull out of it.

Each month has its own *mazal*, or constellation. Shevat is lined up with *Dli*, which is literally a bucket but is more commonly known in English as Aquarius, the Water Bearer. What is the message that *Dli* carries? This line of inquiry becomes particularly difficult when we reflect on how the Jewish People are said to be represented by a *Dli*.[6] Why would the Chosen People be compared to a bucket? This seems arbitrary at best and perhaps even negative—for isn't a bucket only valuable inasmuch as it serves its contents? Why would the nation that validates the creation of the world be called a tool of secondary importance?

Let's follow a second train of thought to the "New Year for the Tree" observed in Shevat. Many are probably familiar with the question regarding the usage of the singular word "tree." A common explanation given is that this month's job relates to *the* tree—the Tree of Knowledge of Good and Evil in Gan Eden. So our second question is, what is the connection between Shevat and the *Eitz Hada'as*? How does it connect to the holy nation compared to the bucket?

Before we attempt to answer these two questions, let's pursue a third line of thought: The *Sefer Yetzirah* writes that each month is aligned with a particular human faculty. In Shevat we are enjoined to rectify *le'itah*. This term has been left untranslated so we can analyze its nuanced meaning.

Le'itah is a type of eating, but if it were just that we would call it *achilah*. *Le'itah* is more accurately described as being fed. This is seen when Eisav demanded that Yaakov pour soup down his throat. He said, "*Ha**lite**ini na min ha'adom ha'adom hazeh*—'**Pour into me** now some of that very red stuff'" (*Bereishis* 25:30). So our third question is, how can one rectify *le'itah* if it's not a human faculty in which one can choose to engage? Why didn't the *Sefer Yetzirah* simply say that this month our job is to perfect the faculty of *achilah*?

---

6. I found this idea in the *Bnei Yissaschar* (*Ma'amar* 1 *L'Chodesh Shevat*), quoted from earlier sources.

In order to answer our three questions let's turn to the *Pri Tzaddik's* thoughts (*Ma'amar L'Chodesh Shevat, os* 3) on Yaakov Avinu and *le'itah*. We find in *Vayeitzei* (28:20) that Yaakov prayed that Hashem should provide his basic needs. Yaakov asked for "bread to eat and clothes to wear." *Pri Tzaddik* asks: Isn't it obvious that bread is to eat and clothes are to wear? Why did Yaakov feel the need to spell out what he would do with the food and clothing that Hashem would give him as if Hashem hadn't already known? Rav Tzadok answers that Yaakov wasn't just asking that the food and clothing come to him from Hashem. Yaakov was also asking that the "to eat" and "to wear" parts also come from Hashem.

A person can feed himself and put on his own clothing. In such a case, all we know is that the one eating or getting dressed desires to have the food eaten or the clothing on. But if somebody else feeds a person (like in a *le'itah* situation), or if someone else clothes his body, then one can draw the conclusion that the party doing the feeding or clothing desires the result. So Yaakov was telling Hashem that he didn't just want food and clothing to indulge in at his *own* whim; he only was going to eat if the desire for the food to be eaten came from Hashem. In simpler words, Yaakov was saying that he would never eat for his own pleasure, but only in situations where it was Hashem's will that he should eat. Similarly, he wouldn't wear clothing for his own gratification, but only if wearing the clothing would conform with Hashem's Will. When Yaakov said he wanted "bread to eat," he meant that he wanted "bread" to come from Hashem as well as the injunction and desire "to eat." In this way, Yaakov told Hashem that if he encountered a situation where there was a conflict of interests because he (Yaakov) wanted to eat or wear clothing but Hashem didn't want him to, he (Yaakov) would bow in submission before the Will of his Creator.

Now perhaps we can backtrack and answer our questions. Let's start with number three. Why is the faculty of Shevat referred to as *le'itah* instead of *achilah*? The nuanced term *le'itah* was chosen so we would learn exactly *how* to rectify eating. Eating is not supposed to be done for our own personal gratification. *Achilah* is merely a self-serving act in which every beast in the animal kingdom engages. A true servant of a king does every act to serve his master. Eating should be done with the intent that Hashem wants us to be healthy. Eating should be done with the intent that we cannot serve our Master without the energy given to us by food. Eating should not be done as an *achilah*, based on our own will, but rather as a *le'itah*, based on Hashem's Will.

This is the connection between Shevat and the Tree of Knowledge. What was the great sin of Adam and Chavah? All they did was eat a fruit! The answer is that they ate because *they* desired it, while the Divine said it should not be eaten. There was a conflict of interest, and the purpose of creation was so Man would earn eternal bliss by learning to sublimate his own will before his Creator's. Tragically, he failed, and we now must rectify this sin by doing what he was supposed to do.

Let's conclude by answering our initial question. What is the significance of the bucket and why are the Jewish people compared to it? The answer is that a bucket's sole purpose is to contain water. Water is a term used to refer to Torah (*Bava Kamma* 17a). Very often, people get caught up in different causes and philosophies and forget that the Jewish people have only one reason for existence—to keep the Torah.[7] Man was made with desires and it is normal for a person to have aspirations and dreams. But our job as Hashem's people is to learn to sublimate our desires before His Will. If we can learn to bow in subservience before Hashem's Will when conflicts of interests arise, we'll succeed in our job in life and become the holy people we are meant to be.

7. See footnote 5.

Adar

Holy Game of Hide-and-Seek

The *Shulchan Aruch* states: *Mishenichnas Adar, marbim b'simchah*—"When the month of Adar comes in, we increase our joy." Now this is a tall order. Happiness is not controlled by a switch that we can flip on or off. How can Hashem expect us to just decide to increase our joy in a particular month? It is obvious that Hashem only commands us to do things that we are capable of doing. How can we work on upping our level of happiness in the month of Adar?

Here's another question about Adar: Every month has a constellation, or *mazal*, that represents the uniqueness of the month over which it rules. The *mazal* of Adar is *Dagim*—Pisces, the Fish. What fish-like characteristics does this month have that it should be exemplified in such a way?

I think we can answer these questions with the following story:

Yaakov, a good and involved father, was playing hide-and-seek with his six-year-old son. "Okay, Shmully, here's how it works. You close your eyes and count to thirty. I'm going to hide and you have to try to find me," explained his father. Shmully scrunched his eyes closed and counted as his father hurriedly hid under the kitchen table. With a call of, "Ready or not, here I come," Shmully started searching for his father. He ran around the living room, his face lit up with a joyful glee as he checked all the usual spots. Not meeting with success in the living room, he eagerly moved on to the kitchen, all the while calling out, "I'm gonna find you, Daddy!" Shmully took a peek under the table and squealed in delight. "I found you!"

Let's take a moment to analyze this common scene. Why is little Shmully so happy when he plays this game? What is he so joyful about as he searches for his father? Don't you think he should be sad and worried? Shouldn't he be thinking, "Where is my Daddy? He's gone! I can't find him!"? We all know the answer. Shmully is not upset because he knows his father is around somewhere. He's not gone; it's just part of the challenge—to try and find him.

If there's one thing we learn from the Purim story, it's about *hester panim*—how Hashem often hides behind a mask and we can't always succeed in seeing Him. But there is no need to be overly distressed. It's all part of the challenge—to try and find Him. *Those who seek Hashem will be joyful of heart* (*Divrei Hayamim I* 16:10). There is no reason to be sad or upset if we can't see Him. We know that He's always there, *peeking out between the cracks* (*Shir Hashirim* 2:9). We are therefore confident that Hashem is always there, silently cheering us on as He eagerly waits for us to find Him.

Now maybe we can connect the month of Adar to fish. The Gemara teaches that fish multiply to such a great extent because *ayin hara* (the evil eye) does not exert any dominion over them. The reason for this is because fish are "hidden from the eye." While land creatures are visible for all to see, fish are concealed beneath the ocean waves.

Is it coming together yet? Throughout Adar, we are expected to focus on the One Who is hidden. Hashem, like a fish, is hidden from sight. Our job, though, is to seek Him out. The *avodah* of Adar is essentially a month-long game of hide-and-seek. We are commanded to increase our joy during this month. We do so by searching for evidence of *hashgachah pratis*—Divine involvement in our lives. If we search for Hashem, we'll find Him—and feel a lot happier.

May we merit succeeding in our holy game of hide-and-seek.

Powerball vs. Pisces

Halachah states that when Adar comes in we must increase our joy. This may not be too hard for many, as Adar's holiday, Purim, holds a dear place in their hearts. However, a more difficult aspect of Adar is the area of spiritual growth. What is the *avodah*, specific job, of Adar?

We know that a month's holidays inform us of our job for its duration. But Purim is very unique. Most Jewish holidays have a fairly similar structure in which a historical narrative is highlighted to serve as the springboard for us to focus on a specific area of growth. Usually, we commemorate narratives of extraordinary occurrences because supernatural events reveal a higher reality that we must incorporate into our own reality. That is to say, the miraculous plagues in Egypt enable us on Pesach to recognize Hashem's mastery over nature, while the Revelation at Sinai creates a yearly renewal of the covenant with the Jewish people on Shavuos. We live in a sukkah for a week in an effort to incorporate the lessons of the Clouds of Glory, and we light the menorah to imbibe the eternal message of Chanukah.

Purim, however, stands out in that it is the only celebratory holiday where there are no supernatural events commemorated. This is for the obvious reason that there *were* no supernatural events that occurred in the Purim story. While as Jews we believe that all the events that took place were orchestrated by Hashem, if one wanted to he could chalk it all up to sinister schemes, clever political machinations, and a hefty dose of good luck. So what higher reality is revealed in the story of Purim? All other holidays recall occurrences that contain incontrovertible evidence of G-d's Hand, so it is beneficial to publicize the miracle. But if His Hand is hidden, then what can be gained by re-telling the story?

Before attempting to answer this question, let's add another. A large portion of the Megillah centers around Haman HaAgagi. This Persian prime minister was not your run-of-the-mill Jew-hater. Haman was a member of the nation of Amalek, a people the Torah labels as the archnemesis of the Jewish people. In addition to this ignoble status, the Torah commands that Amalek must be wiped off the face of the Earth with nary a remnant left behind. Why does the Torah take such a strong stance regarding this nation? Historically, many nations have tried to do away with us. Yet, there is no mitzvah to eradicate the Egyptians, Babylonians or Greeks. What is it about Amalek that is so intolerable?

Let's add one more question before we move on to provide some answers. The *mazal* of Adar is *Dagim*, which literally translates as "Fish" (but is more commonly known as Pisces). What is the relationship between Adar and Pisces? In what way does Purim connect to fish?

In order to answer these questions, we'll need to take a detour and analyze the timeline of the Megillah. The verses do not clearly spell out all the dates of the events that occurred. This may lead one to believe that the events in each chapter were immediately followed by the events of the following chapter. However, this understanding would be inaccurate. The events of the Megillah were actually spread out over a span of close to ten years, with the feast occurring in the third year of Achashverosh's reign and Esther's appointment taking place in the seventh. Mordechai foiled the assassination plot after Esther had already been settled in the palace, while Haman's plot to kill the Jews took place in the twelfth year of Achashverosh's reign! Why is the Megillah written in an obscure way that allows for the impression that the events happened in quick succession if that wasn't truly what occurred?

What makes this question even starker is the first verse in the third chapter of the Megillah: *Achar hadevarim ha'eileh...*—"After these events occurred (Achashverosh [proceeded] to elevate Haman)..." In the Holy Tongue there are no synonyms. Each word has a precise connotation. The words *achar* and *acharei* both mean "after." However, *achar* means "after a short period of time" while *acharei* means "after a long period of time." Why does this chapter start with the word *achar*—implying that these events took place shortly after those of the previous chapter, when in reality they took place a few years later?

The Megillah wasn't written in this way by accident. On the contrary, this wording was used *precisely because the authors wanted to communicate the message that there was a cause-and-effect sequence.* While an ordinary person may not have been able to see the

Guiding Hand in the events, Mordechai and Esther were able to connect the dots. To them, it was eminently clear (at least in retrospect) that Hashem arranged the feast in order to get rid of Vashti and clear the way for Esther. Mordechai Hatzaddik was able to recognize that Hashem didn't allow Haman to rise to power until Achashverosh was first indebted in some way to Mordechai. That is why the word *achar* is used—to show how the sickness (Haman's elevation) was immediately preceded by the cure (Mordechai saving the king's life). The Megillah was written in such a way that one could clearly see how, from a Divine perspective, many disparate events that occurred over the span of a decade were really interconnected and how they were being orchestrated by the ultimate Puppeteer.

This idea is the higher reality that was revealed on Purim. True, there were no open miracles on Purim. But when Mordechai and Esther managed to connect the dots and recognize Hashem's guiding Hand, they seized the moment and wrote *Megillas Esther*—a concrete description of how Hashem truly *is* involved in our lives.

Amalek, on the other hand, embodies the antithesis of this worldview. While we as Jews ascribe *hashgachah* (Divine Providence) to every step of our lives, the Amalekite philosophy allows for the possibility of randomness and coincidence. The Chumash and the Megillah both refer to Amalek with a variation of the word *kar*, which means "happenstance." The Amalekites believe that Hashem, while He may exist, is not necessarily always watching over every little thing that occurs. They believe that sometimes Hashem is not paying such close attention to the world. They allow for the belief that in certain months (like Adar, the month during which Moshe died) Hashem is not in complete and total control.

This is an idea that we cannot tolerate. True, at times Hashem is more hidden. But He is always there. The Torah doesn't demand that everyone be Jewish; Jews and gentiles both can form a relationship with the Almighty and serve Him. That is why the world was created. But if a nation denies (or is unwilling to fulfill) the sole purpose of the planet, that nation has no place on it.

Now we can understand why the *mazal* of Adar is fish. Fish are unique creatures in that they have no eyelids. Their eyes are always open. The Shelah Hakadosh explains that this can allude to the *pasuk* in *Tehillim* (121:5): *...the Guardian of Israel neither sleeps nor slumbers.* His eyes are always open, watching us and guiding us every step of the way. This also can explain why the holiday is called *Purim*, which means "lots." Although the lottery that Haman drew may seem to have been a very minor detail in the story, in reality it is a representation of the eternal struggle between our idea of *hashgachah* versus Haman's idea of chance. Powerball versus Pisces. Are all things random or part of a master plan?

May we merit opening our eyes and seeing G-d's guiding Hand in every step of our lives.

Holidays and Loose Pockets

Every month has its own unique *avodah* (job). With that in mind, let's begin our analysis of Adar. What is the area in which we must grow during these days? In order to answer this question, we need to take an extended journey through the mitzvos of Purim. Only an understanding of the commandments of the prime day of the month will enable us to understand the month in its entirety.

Mordechai and the Sages enacted four distinct mitzvos to observe on Purim in celebration of the great salvation Hashem wrought for us in their days. First, we must read the Megillah. This is a logical way to recall the story and give thanks to Hashem for saving us from Haman's terrible decree. We also were given a mitzvah to eat a festive holiday meal, to rejoice and celebrate the salvation. So far, so good. However, the next two mitzvos are a bit more enigmatic. They require us to give out *mishloach manos*—gifts of food to our friends, and *matanos l'evyonim*—monetary gifts to those in difficult financial straits. Why do you suppose the Sages wanted us to fulfill these two injunctions on Purim? What do they have to do with Mordechai, Haman, and all that Hashem did for us?

The Malbim writes in his commentary on the Megillah (9:22): *Why were we commanded to rejoice with eating the meal and also with giving gifts? This is because the verse informs us that the month was transformed in two ways—from "groaning to rejoicing" and from "mourning to a holiday." The rejoicing with eating was commanded as a parallel to the groaning-turned-rejoicing, and the gift giving is a parallel to the mourning-turned-holiday.*

It seems that the Malbim is explaining that the mitzvos of *mishloach manos* and *matanos l'evyonim* were enacted because the Sages were trying to create a holiday, and gift giving is (apparently) the best way to make an otherwise regular day into a holiday. This is hard to understand. In what way does gift giving make a regular day into a Yom Tov?

The Malbim goes on to explain how this works: *The whole purpose of any holiday is for one to separate from the vanities of the world and [cling] to Hashem and His Torah. Mordechai and the Sages wanted to make a new holiday, but they couldn't—because only the Torah has the power to create a holiday. That is why they enacted these mitzvos of gift giving—because this too is a mitzvah.*

It seems that the Malbim is saying that the Sages wanted to make a holiday, but they didn't have the "right" or "ability" to do so. Therefore, they decided to enact a requirement to fulfill two of the mitzvos that would engender the same results (i.e. closeness to Hashem) as a real Biblical Yom Tov.

However, this Malbim needs explanation. First of all, why were these two gift-giving mitzvos chosen to help bring us closer to Hashem on the Rabbinic holiday? Don't *all* the mitzvos bring a person closer to Hashem? Why did the Sages choose this specific type of mitzvah to turn Purim into a spiritual day? Secondly, if gift giving is indeed the best way to turn a mundane day into a Yom Tov, why did the Sages enact *mishloach manos* and *matanos l'evyonim* on Purim but not on Chanukah?

It seems obvious that *these* mitzvos were chosen for *this* day specifically because they are related to the overall theme of Purim. Let's explain how this is true. We are all undoubtedly familiar with the Megillah's protagonists, Mordechai and Haman. However, in a broader sense, the Purim saga was not so much about these two men as it was about their nations and philosophies. Mordechai is the representative of the Jewish people, while Haman is the archetype Amalekite. What are these two nations all about? The midrash (*Esther Rabbah* 8:5) and Rashi (*Ki Seitzei* 25:18) identify the Amalekite worldview with one simple word: *mikreh*—coincidence. The Amalekites profess that all that occurs is merely a happenstance and coincidence. The Jewish people, on the other hand, know that everything that occurs is the result of Hashem's guiding Hand.

If you'll join me, I'd like to paint a picture of how one would (and should) live if he subscribes to the Amalekite worldview. He should constantly be concerned with himself. He should hoard all that he can get his hands on and stash it away in many safe places, because one never knows when a natural disaster may occur and cause him to lose his assets. A responsible adherent to the Amalekite philosophy cannot afford to share what he has with others because he may need it himself one day. Additionally, one could make the socio-ethical argument that it is downright wrong to share because it encourages dependency. The rule of the Survival of the Fittest should dictate that only the toughest make it in life.

A Jew, on the other hand, can share all that he has with others because he knows that it will not negatively impact his own financial standing. A Jew knows that Hashem is taking care of him, and there is therefore no urgent need to prepare for every and any eventuality. All a Jew needs to do is prepare for his basic needs, and then he can share all that he has with others.

Now perhaps we can explain the Malbim's idea that gift giving can turn Purim into a holy day. When a person loosens his pockets and gives to others with abandon, he is taking the steps to separate from the vain concerns of the world and the need to worry about what may occur in the future. By giving from what he has, he learns to cling to Hashem in faith and trust. *Mishloach manos* and *matanos l'evyonim* help a person imbibe the idea that money need not be hoarded, because Hashem can provide for us. In other words, gift giving is the best way to turn Purim (as opposed to Chanukah) into a holiday, because it trains one to be a Jew and not an Amalekite.

This is an area in which we can grow in Adar. Let's use the whole month to focus on thinking and acting like Jews. Let's realize that everything we get is from Hashem—because there is no happenstance, only Divine Providence. Let's be more giving because we know that we can; Hashem can replenish whatever we disburse.

Adar Rishon: The Hidden Fish

Ostensibly, it is difficult to extract lessons from Adar during a leap year due to the numerous disagreements as to how we understand the extra month. On the one hand, the Gemara (*Megillah* 6b) states that, *there is no difference between Adar Rishon and Adar Sheini.* Then again, Purim is only truly celebrated in Adar Sheini. An additional problem is that our modus operandi of aligning the tribes with the months to discover the monthly *avodah* is problematic, being that the tribal alignment is hotly debated with regard to the extra Adar. Some commentators place Naftali as the tribe of both Adars. Others align Naftali with the regular Adar but insert Levi into the extra Adar (*Tzemach Tzedek*), while yet others (*Kedushas Levi*) match Yosef up with Adar so his line can be split into Ephraim and Menashe during a leap year. There is even an opinion that there is no tribe matched up with Adar Rishon altogether!

I think the best way to go about this is to turn to the *mazal* (constellation) of the month, *Dagim*—Pisces, the Fish.[1] This

1. Then again, there are also those who maintain that the extra Adar has no *mazal*. See *Bnei Yissaschar* (*Ma'amar* 1 *L'Chodesh Adar, os* 11). Also see Chida

mazal is different than the others in that it is plural, while the others are described in the singular form—the Ram, the Bull, the Crab, the Lion, etc. Although the Twins and the Scales are also plural, that is only because they are plural by definition. There is no such thing as one scale or one twin. *Dagim* is the only *mazal* in the calendar that is "unnecessarily" plural. It would therefore seem that *Dagim* is the *mazal* of both Adar Rishon and Adar Sheini.

What makes fish different than all other creatures? They definitely have many unique qualities, such as the fact that they swim, live in the water, etc. But in the Torah's view, they are significant for an altogether different reason. The Torah describes fish as being unique because they are covered over by water and hidden from the eye. The Gemara (*Brachos* 20a) describes how Yaakov blessed Yosef that his children should be like fish, and how Yosef's descendants are always protected from *ayin hara* (evil eye) as a result. The Gemara explains that fish are protected from *ayin hara* specifically because they have that quality of being covered and hidden. (Incidentally, this would support the view that Adar is aligned with Yosef.)

But what *avodah* is there in hiddenness? Perhaps one will suggest *tznius*, modesty. This is definitely a Torah ideal, as we see from the Mishkan. The Mishkan exemplified modesty, for the gold-plated staves were covered by tapestries, and the beautiful tapestries were in turn covered by goat skins. This bespeaks the idea that just because something is beautiful doesn't mean that it needs to be shown off to everyone else. We also see a support for this ideal from the verse in *Michah* (6:8) where we are charged: *...walk modestly with your G-d.* An added dimension to the verse can be found in Rav Shimon Schwab's explanation that the word "with" is specifically used to emphasize that Hashem Himself is modest in His ways. A third source for the ideal of hiddenness can be found in the Maharal (*Nesivos Olam, chelek beis, Nesiv Hatznius*) that the salvation of Purim was only

---

(*Devash L'fi*). However, we will work with the assumption of many great *sefarim* that it indeed has the *mazal* of *Dagim*.

through the merit of Esther's *tznius*, which she inherited from her forebearer Shaul, who inherited it from his forebearer Rochel. So perhaps the work of Adar is to be hidden and not put all our blessings on display.

On the other hand, one could say that hiddenness is not such a good thing. The Gemara (*Chullin* 131b) describes how the Purim story had no open miracles because the Jewish people were being subjected to the curse of *hester panim*—Hashem hiding His Face. So is hiddenness a good quality or not? Would Hashem reveal His Presence in an ideal world or modestly hide it? How can we resolve this contradiction?

In order to answer this question, let's turn to a Rashi in *Ki Sisa* (*Shemos* 34:3). Rashi is explaining why Hashem did not make a big ceremony for the second set of *Luchos* (*And no man may ascend with you…*) like He did for the first set. Rashi states: *The first Tablets were given with fanfare, great sounds, and amongst the entire congregation. Consequently, they were inflicted with* ayin hara *(evil eye). [Therefore, the second* Luchos *were given in a private fashion, for] there is nothing better than* tznius. This Rashi is quite perplexing. How could we dare say that Hashem did something wrong and immodest the first time around? Also, who gave the *ayin hara*?

Let's take a detour to understand what *ayin hara* is and then perhaps we can answer these questions. I heard quoted that Rav Dessler writes in the *Michtav M'Eliyahu* that *ayin hara* is an ethical issue. Hashem may bless you with all sorts of wonderful things, but you shouldn't be sticking it in other people's faces. If what you have causes pain to other people, Hashem may decide you shouldn't have it anymore. Therefore, the best option is to keep your blessings hidden.

Now perhaps we can explain what happened at Har Sinai. Of course, Hashem did nothing "wrong" by giving the *Luchos* with all the pomp, ceremony, and fanfare. The Handiwork of G-d was given with all the splendor it deserved. However, the great revelation provided a test. Perhaps it was a test for the gentile nations, who failed and cast a jealous evil eye toward the Chosen People. Or perhaps it was a test for the Jewish people themselves who, upon finding it difficult to fully submit to Hashem's service, cast an *ayin hara* toward the *Luchos* that contained the Law that would control their lives from that moment on. Either way, the great fanfare caused pain, which engendered an *ayin hara* to be cast on the *Luchos*. That *ayin hara* consequently caused the *Luchos* to be broken. Hashem saw that, based on the level of the ones being tested, it was necessary for the second set to be given privately.

Now let's backtrack to our discussion about hiddenness. Of course, *tznius* is a laudable trait. And of course, it's not a good thing for Hashem to be hidden. But the differentiation we need to make is between revealed and ostentatious. Nobody ever said that everything beautiful has to be locked away. But nobody said that the blessings you have should be put in other people's faces either. The Mishkan was beautiful but was covered over so it wouldn't be ostentatious. Hashem surely did the right thing at Har Sinai because, ideally, He should be revealed in the world. But the grandiose display was challenging for people. Hashem tested His creations and, tragically, they failed the test. And, as a final point, the thousands of beautiful species of fish in the sea are hidden from the eye and protected from *ayin hara* specifically because they never put themselves in other people's faces. So let's use Adar as the time to work on the lesson of the hidden fish and become truly modest.

Adar Sheini: Monkeys, Amalek, and Gigantic Seesaws

In years during which there is an Adar Sheini, we have a double-header opportunity to learn about the *avodah*, specific job, of Adar. The idea of two months of Adar is complex, and I don't know enough to discuss it in any meaningful way. However, since *Dagim* is a plural word, the commentators explain that Pisces, Fish, is the *mazal* for both Adar Rishon and Adar Sheini. This indicates that we should grow in a fish-related way in both months of Adar. Let's see if we can accomplish this goal.

We begin our analysis of Adar with a look at the mitzvos of the month. Purim is celebrated with four distinct observances: *mishloach manos* (giving gifts of food to friends), *matanos l'evyonim* (gifts to the poor), *mishteh* (a festive meal), and *krias haMegillah* (reading the Megillah). Let's analyze this last one. On no other holiday is there a mitzvah to take a distinct scroll and read over the story behind the holiday. Even on Pesach, when we are indeed commanded to recount the Exodus, we don't have to recite specific words. The fact that the Purim saga requires its own text gives this mitzvah more of a formalized air. Why was it set up in such a way?

Another mitzvah observed in Adar is *zechiras Amalek* (remembering Amalek). This mitzvah entails reciting the *pesukim* that recount the episode of how Amalek attacked us as we left Egypt and reiterating the obligation to wipe out this evil nation. Actually, *Zachor* is a mitzvah all year round, and is even one of the six daily mitzvos. However, our Sages set the Shabbos before Purim as the specific day that we re-read the account. The Shabbos before Purim is a logical choice, as Purim also recalls how an Amalekite tried to do away with us. But let's analyze this mitzvah. How can we fulfill it nowadays? The ancient nation of Amalek (like all nations of old) has mingled with all the other peoples of the Earth and is no longer distinguishable! Another fair question one may ask is, what will happen when Amalek is indeed finally eradicated? Will the 558th commandment become obsolete? Will the eternal Torah then only have 612 mitzvos?

Before addressing these questions, let's ask one more. Each month is associated with a particular letter, with Adar aligned with *kuf.* All letters have a "short" form in which the letter is made up of a single letter, as well as a "full" form in which the letter is spelled phonetically. The full form of *kuf* is *kuf vav pey.* Thus, since the holy tongue has no intrinsic vowelization and all variations are acceptable, *Pri Tzaddik* (*Ma'amar L'Chodesh Adar, os*

3) explains that it can be said that Adar is associated with *kof*—a monkey. What does Adar have to do with monkeys?

The *Pri Tzaddik* suggests an explanation based on a brilliant homiletical insight into the Gemara in *Bava Basra* (58a): *As a monkey to a man, so too a man to the* Shechinah (Hashem's Divine Presence). The simple understanding of this Gemara is that the spiritual radiance of man is nil compared to the *Shechinah*, just as a monkey's spiritual radiance cannot compare to that of a human being. Rav Tzadok suggests, however, that this Gemara is alluding to the 608th mitzvah—to copy Hashem's ways. Just as a monkey mimics man, man should mimic the *Shechinah*. However, this is a mitzvah all year round. If the *kof* is associated with Adar, there must be a specific Adar-related manner in which we should mimic Hashem during this time. But what is it?

Let's start by answering this last question and then backtrack to the other two. The midrash on the Megillah (*Esther Rabbah* 3:12) describes how Haman tells Achashverosh that it's safe to make a move on the Jews because "*their G-d is sleeping*." The midrash records that Hashem exclaimed, "You think I'm sleeping? Behold! The Guardian of Israel never sleeps nor slumbers!" You see, Haman made a fatal error typical of his nation, Amalek. He assumed that everything that occurs is a coincidence, that Hashem is not always involved. He reasoned that if the Jewish Temple was not rebuilt, it must be because G-d is sleeping. He supposed that the world simply moves along and things just occur as they happen to unfold. Haman probably assumed that as far as anyone knows, the world came from a random Big Bang and we all evolved from monkeys!

Foolish Haman didn't realize that Hashem didn't abandon the world. Rather, Hashem is like a fish. Fish don't have eyelids so their eyes are always open. They might not be seen, but they can always see. Amalek believes that Hashem's silence indicates His absence. Jews, on the other hand, believe that everything that happens is occurring under the watchful eye of the Almighty Creator.

This debate has great consequences far beyond the philosophical beliefs of these respective nations. When the progenitors of these two nations were in the womb together, the prophet declared, *"When this one rises, the other shall fall"* (*Rashi, Bereishis* 25:23). Amalek draws lifeforce and power from the belief in coincidence. Jews draw their power from the belief in the Divine Providence of G-d. The Torah describes how in the original battle with Amalek, *When Moshe would lift his hands, Israel would be strong. When Moshe would lower his hands, Amalek would overcome* (*Shemos* 17:11). Rashi rhetorically asks: *Did the hands of Moshe win the war? [Of course not!] Rather, when Israel would raise their eyes to Heaven and subject their heart to their Father in Heaven they would prevail. But when they would not, they would fail.* Neither can live while the other survives.

Now let's tie it all together and answer the other questions. Like the eyes of a fish, Hashem's "Eyes" are always open, for the Guardian of Israel neither sleeps nor slumbers (Shelah Hakadosh). And just as a monkey mimics those he sees, we must mimic this quality. We should keep our eyes open and always seek Hashem out. He is not visible (just as fish are hidden from the eye), but we are obligated to search for evidence of Divine involvement. Additionally, we must learn that *hashgachah pratis* (Divine Providence) stories are not simply good bedtime reading material. They are the lifeforce of Klal Yisrael. They are the only way we can weaken the power of Amalek. By strengthening our recognition of Hashem's involvement in the world we are fulfilling the injunction to eradicate Amalek—even in our days! This is a commandment that is truly eternal, for there are infinite levels of faith to which we can grow. Therefore, even once Amalek is physically annihilated, we will still be able to fulfill the mitzvah by continuing to strengthen our belief in Hashem's constant involvement in our lives.

As a final point, we can suggest that this is why there is a formalized mitzvah to read the Purim story. If Adar is the time for *hashgachah pratis* stories, then what better proof of Divine Providence could there be than the Megillah—in which the disparate events over the course of a decade are in retrospect proven to be part of a Divine Master Plan?

Our job is now laid out for us. To use a humorous imagery,

envision a gigantic, world-straddling seesaw with two monkeys on it. One monkey is a highly evolved, coincidence-oriented Amalekite monkey, and on the other side is a Jewish monkey who tries to mimic the *Shechinah*. The two are struggling for supremacy. This struggle transcends time and space. Neither can live while the other survives. When this one rises, the other shall fall.

However, the Gemara (*Bava Basra* 123b) says that Amalek will only fall by the descendants of Yosef—the tribe compared to a fish. Let's grow in this regard and thereby merit the day when the whole world will see that Hashem is One and His Name is One.

Purim: Concealing and Revealing

Every Jewish holiday requires preparation beforehand. For Chanukah, one must buy oil. For Sukkos, one must build a sukkah. For Pesach, well, I'm not even going to try describing what one must do for Pesach. What about Purim? What is the buzz that one hears as Purim approaches? In my area (and I assume most Jewish communities in the world), it's about two things: costumes and *mishloach manos* (gifts of food for friends). All the children (and many adults) are fretting about what they are going to dress up as, and many families are working on some sort of a "theme" for their *mishloach manos* packages. If these two ideas are so central to the holiday, we should invest a little bit of time into discovering their meaning and how they connect to Purim.

Let's start with costumes. Why do we dress up on Purim? It is not a halachah or a mitzvah, so from where did this custom originate? The Gemara in *Chullin* (131b) is often quoted to explain this tradition: *From where can one find an allusion to Esther [and the Purim story] in the [Five Books of the] Torah? The answer is, it says (*Devarim *31:18):* Va'anochi hasteir asteir panai bayom hahu—*"I will surely have concealed My Face on that day."*

This Talmudic passage is explaining that the name "Esther" is cognate with the Hebrew word for "hidden"—*hasteir.* Indeed, throughout the entire Purim saga, G-d performed no open miracles and only saved His people through hidden, natural means. It seems that the custom of wearing masks (or dressing up, i.e. masquerading) stems from a form of *Imitatio Dei*—copying the ways of G-d. If Purim is the time that Hashem hid His Face, then it is the time that we too should hide our faces.

But this needs explanation. What do we gain by imitating Hashem in this fashion? It would indeed be logical if we were imitating kind acts that Hashem does: *As He dresses the unclothed, you dress the unclothed. As He visits the sick, you visit the sick…* (*Sotah* 14a). It would be similarly understandable if we were recreating a miraculous event (by dwelling in a sukkah, lighting a menorah, or eating matzah), in order to imbibe the message of G-d's might. But what can we gain from recreating something as negative as G-d's absence? Would it not be more appropriate to celebrate the deliverance instead of the sickness that preceded it?

Before addressing this question, let's analyze the mitzvah of *mishloach manos.* Everyone enjoys giving gifts. The Torah even endorses it. Is there any better expression of *v'ahavta l'rei'acha kamocha* (love your neighbor as yourself) than this? But why did our Sages see fit to institute such an expression of love and unity specifically on Purim? What is it about Purim that necessitates working on unity?

In order to answer these questions, we need to understand the idea of *hester panim.* What does it mean that G-d will "hide His Face"? Does Hashem ever go away and not pay attention to us? Heaven forbid! Hashem is our loving Father and He never takes His eyes off us for a moment. However, sometimes He's concealed. That means that He *is* there, but we can't *see* Him. We know that Hashem is benevolent, but sometimes it seems our life is being controlled by something more severe. If only we could recognize that whenever we see that harsh countenance, the difficulties are in truth coming from a benevolent Source! If we could somehow learn this lesson, wouldn't life be much easier?

Perhaps now we can suggest an answer for our first question.

The custom of masquerading on Purim comes not from an imitation of Hashem's concealment on Purim, but rather from the recognition that the travails of Purim were only the results of a mask. Meaning, in the times of Mordechai and Esther, we had one of those rare moments when we were able to discern that behind all the hardships was a loving Father who was orchestrating everything solely for our benefit. We were able to realize that the harsh countenance that we had perceived was not Hashem's *true* Face, but just a mask. This explains why masks are truly something to celebrate.

Now let's move on to answer the second question. There is a famous quote from the *Zohar*: *Yisrael v'Oraisa v'Kudsha Brich Hu chad hu.* This extremely esoteric idea means that there is a special connection between Hashem and the Jewish people (and the Torah) in which qualities found in one will naturally be found in the other.

Based on this idea, I believe we can answer our second question. Why is it so important to work on unity between Jews on Purim? Because unity amongst Jews creates unity Above. Sometimes we have difficulty perceiving that everything is coming from our benevolent Father. Sometimes things look like they are coming from a second, less benevolent source. If we were able to recognize that everything is coming from the same one loving Father, we wouldn't find life's problems so overwhelming. That's why on Purim we go about giving gifts. Gifts increase love, brotherhood, and unity below. And when there is unity below, there is a ripple effect that (in some inscrutable and abstruse way) causes unity Above.[2]

I think that by now we should have a clearer picture of what Purim is all about. In exile, Hashem is often concealed—*I will*

---

2. I don't recall where this is stated explicitly, but perhaps an allusion to this idea can be found in the *Pri Tzaddik* (*Ma'amar L'Purim, os* 3 and 4) where he says: *The root (i.e. purpose) of the mitzvah of* mishloach manos *is to unite the hearts of the people to be* k'ish echad b'lev echad—*like one man with one heart...and in these days Hashem also fulfills the mitzvah of* mishloach manos *by uniting His Name (i.e., the* Yud *and* Hey *with the* Vav *and* Hey*...).*

surely have concealed My Face on that day. But there are incidents that occur once in a while (like the Purim story) that teach us that hardships are only a mask and not Hashem's true "Face." We must focus on that idea and then the next step is to remove the problem that engendered the dual perception in the first place. Through *mishloach manos* (*and matanos l'evyonim*—gifts for the poor), we spread love and unity to thereby create the unity Above that will allow us to realize that everything is indeed from—and solely from—His loving Hand.

Appendix

	Mazal	Letter	*Pe'ulah* (Faculty)	Tribe (according to Arizal and Gra
Nissan	*Taleh*/Aries, the Ram	*hey*	*sichah*	Yehudah
Iyar	*Shor*/Taurus, the Bull	*vav*	*hirhur*	Yissachar
Sivan	*Te'umim*/ Gemini, the Twins	*zayin*	*hiluch*	Zevulun
Tammuz	*Sartan*/Cancer, the Crab	*ches*	*re'iyah*	Reuven
Av	*Aryeh*/Leo, the Lion	*tes*	*shmi'ah*	Shimon
Elul	*Besulah*/Virgo, the Virgin	*yud*	*ma'aseh*	Gad
Tishrei	*Moznayim*/ Libra, the Scales	*lamed*	*tashmish*	Ephraim
Cheshvan	*Akrav*/Scorpio, the Scorpion	*nun*	*rei'ach*	Menashe
Kislev	*Keshes*/ Sagittarius, the Archer	*samach*	*shinah*	Binyamin
Teves	*Gedi*/ Capricorn, the Goat	*ayin*	*rogez*	Dan
Shevat	*Dli*/Aquarius, the Water Bearer	*tzaddi*	*le'itah*	Asher
Adar[4]	*Dagim*/Pisces, the Fish	*kuf*	*schok*	Naftali

Manhigim 'nefesh Zachar U'nekeivah	*Tzirufei Shem*[2]	*Eivrei Harosh*
'egel yemin (right foot)	YKVK	*galgalta* (skull) *(nukva)*
ılya yemanis ·ight kidney)	YKKV	*ozen* (ear) *(nukva)*
regel smol (left foot)	YVKK	*ozen* (ear) *(nukva)*
yad yemin (right hand)	KVKY	*ayin* (eye) *(nukva)*
ulya smolis left kidney)	KVYK	*ayin* (eye) *(nukva)*
ad smol (left hand)	KKVY	*chotem* (nose) *(nukva)*
mara gallbladder)	VKYK	*galgeles* (skull) *(zachar)*
łakin (small intestine)	VKKY	*ozen* (ear) *(zachar)*
keivah[3]	VYKK	*ozen* (ear) *(zachar)*
aveid (liver)	KYKV	*ayin* (eye) *(zachar)*
kurkiban	KYVK	*ayin* (eye) *(zachar)*
chol (spleen)	KKYV	*chotem* (nose) *(zachar)*

1. There are alternative opinions. The Raavad matches the tribes from Nissan to Adar: Reuven, Shimon, Levi, Yehudah, Yissachar, Zevulun, Binyamin, Dan, Naftali, Gad, Asher, Yosef.

   The *Midrash Peliah* matches the tribes from Nissan to Adar: Yehudah, Reuven, Ephraim, Dan, Yissachar, Shimon, Menashe, Asher, Zevulun, Gad, Binyamin, Naftali.

   The above chart follows the Arizal and Gra, whose opinions, I believe, are the most conventional.

2. *Yud*=Y, *Hey*=K, *Vav*=V, *Hey*=K

3. *Keivah* and *kurkiban* are both parts related to the stomach that do not have exact translations into English that I know of.

4. Adar Beis is a whole separate discussion beyond the scope of this *sefer*.

Glossary

Achdus—unity

Achilah—eating

Achilas matzah—eating of matzah (commandment at Pesach Seder)

Anshei Knesses Hagedolah—lit. Men of the Great Assembly; institution of Torah Sages that led the Jewish people at the beginning of the Second Temple era

Aravah, pl. aravos—willow(s), one of the Four Species waved together on Sukkos

Arba'ah minim—the Four Species that are held and waved together on the festival of Sukkos

Aveirah, pl. aveiros—sin(s)

Avodah—specific job, work; service

Avodah zarah—idol worship

Avodas Hashem—service of G-d

Ayin hara—evil eye

Ba'alei mussar—leaders of the *mussar* movement (movement that developed in nineteenth-century Lithuania to further ethical and spiritual discipline)

Baruch Hashem—lit. blessed is G-d; Hebrew expression used in place of "thank G-d"

Beis Hamikdash—Holy Temple

Bitachon—trust (in G-d)

Brachah, pl. brachos—blessing(s)

Chadash—lit. new; grain that has been harvested from the new crop prior to the *Omer* offering brought on the second day of Pesach, which may not be eaten

Chag—festival

Chametz—leaven

Chazal—acronym for *Chachameinu Zichronam L'vrachah*, meaning "our Sages of blessed memory"; refers to all Jewish Sages from the Second Temple era until the sixth century CE

Daven, davening (Y)—pray, prayers

Esrog—citron, one of the Four Species waved together on Sukkos

Gematria, pl. gematrios—numerical value(s)

Hadas, pl. hadassim—myrtle branch(es), one of the Four Species waved together on Sukkos

Haggadah—text read at the Seder containing the story of the Exodus from Egypt and all other parts of the Pesach Seder

Halachah, pl. halachos—Jewish law(s)

Hashgachah pratis—Divine Providence

Im yirtzeh Hashem—if Hashem wills it

Kabbalas HaTorah—Acceptance of the Torah

Kavanah—intent

Kippah—skullcap (worn by Jewish males)

Kohen, pl. kohanim—priest(s); males descended from Aharon the High Priest who are allowed to serve in the Temple

Korban—sacrificial offering

Korech—sandwich of matzah and bitter herbs eaten at the Seder

Lashon hara—slander

Limud—learning

Luchos—Tablets containing the Ten Commandments

Lulav—date palm frond, one of the Four Species waved together on Sukkos

Machzor—prayer book used on the festivals and High Holy Days

Mah Nishtanah—four questions asked at the Seder

Mann—manna; food that fell from the sky daily while the Jews wandered in the desert

Maror—bitter herbs eaten at the Seder

Matanos l'evyonim—gifts to the poor (given on Purim)

Mattan Torah—the Giving of the Torah

Matzah—unleavened bread eaten on Pesach

Mazal, pl. mazalos—sign(s) of the Zodiac, dominant constellation of the Zodiac in a given month

Megillah—lit. scroll; often used to refer to the Book of Esther, which is traditionally read from a scroll on Purim

Middah, pl. middos—character trait(s)

Mishkan—Tabernacle

Mishloach manos—gifts of food sent to friends on Purim

Mitzvah, pl. mitzvos—commandment(s)

Navi—prophet

Nes—miracle

Parshah, pl. parshiyos—weekly Torah portion(s)

Pasuk—verse

Pe'ulah, pl. pe'ulos—faculty(ies)

Perek—chapter

Rasha—evil person

Rebbi—teacher (of Torah)

Seder—fifteen-step ritual followed on the first night of Pesach to recall the Exodus from Egypt

Sefer Torah—Torah scroll

Sefer, pl. sefarim—book(s)

Sefirah, Sefiras Ha'omer—seven-week period from the second day of Pesach until Shavuos during which we count daily toward the Receiving of the Torah

Shalom—peace

Shechinah—G-d's Divine Presence

Shema—essential prayer declaring G-d's oneness that is recited by Jews twice daily

Shemoneh Esrei—nineteen-blessing prayer recited three times each day

Shofar—ram's horn blown on Rosh Hashanah

Sichah—speech

Simchah—joy, rejoicing

Sinas chinam—baseless hatred

Sippur Yetzias Mitzrayim—the telling over of the story of the Exodus from Egypt (commandment at the Pesach Seder)

Sukkah—booth in which Jews are commanded to live during the Sukkos festival

Tallis—prayer shawl

Tefillin—phylacteries

Teshuvah—repentance

Toras Chaim—Living Torah

Tzaddik, pl. tzaddikim—righteous man/men

Tzitzis—specially knotted fringes placed on four-cornered garments worn by Jewish males

Tznius—modesty

Yamim Nora'im—Days of Awe, i.e. Rosh Hashanah and Yom Kippur

Yashan—lit. old; status of grain from the new crop after the *Omer* sacrifice is brought on the second day of Pesach

Yasher koach—expression loosely translated as "Good job and thanks!"

Yeshivah, pl. yeshivos—Jewish institution(s) of learning

Yetzer hara—evil inclination

Yetzias Mitzrayim—Exodus from Egypt

Yidden (Y)—Jews

Yom Hadin—Day of Judgment

Zman—time